Nelson
Maths

5

Pupil Book

Karen Morrison
Lisa Greenstein

OXFORD
UNIVERSITY PRESS

OXFORD
UNIVERSITY PRESS

Great Clarendon Street, Oxford, OX2 6DP, United Kingdom

Oxford University Press is a department of the University of Oxford.

It furthers the University's objective of excellence in research, scholarship, and education by publishing worldwide. Oxford is a registered trade mark of Oxford University Press in the UK and in certain other countries.

First published 2022

British Library Cataloguing in Publication Data

Data available

ISBN: 978-1-382-01006-1

1 3 5 7 9 10 8 6 4 2

Paper used in the production of this book is a natural, recyclable product made from wood grown in sustainable forests. The manufacturing process conforms to the environmental regulations of the country of origin.

Printed in Great Britain by Bell and Bain Ltd, Glasgow

Acknowledgements

The publisher and authors would like to thank the following for permission to use photographs and other copyright material:

Cover: Matthieu Nivesse. **Photos: p6:** Laura Dwight / Alamy Stock Photo; **p8:** Andres Felipe Perez/Shutterstock; **p15:** TA BLUE Capture/Shutterstock; **p18:** Bruce Amos/Shutterstock; **p40:** BAZA Production/Shutterstock; **p45:** The Natural History Museum / Alamy Stock Photo; **p49:** James McDowall/Shutterstock; **p50:** CapturePB/Shutterstock; **p51:** Andrew Makedonski/Shutterstock; **p54:** Celso Pupo/Shutterstock; **p58(l):** Suwi19/Shutterstock; **p58(r):** Anna Ismagilova/Shutterstock; **p60:** wavebreakmedia/Shutterstock; **p61:** kocnsiri boonnak/Shutterstock; **p70(br):** Aaronejbull87/Shutterstock; **p70(bl):** Andrea Izzotti/Shutterstock; **p71:** viewimage/Shutterstock; **p81:** Claudio Divizia/Shutterstock; **p92:** photoff/Shutterstock; **p95:** Elena Zajchikova/Shutterstock; **p119:** Africa Studio/Shutterstock; **p121(tr):** Patricia Dulasi/Shutterstock; **p121(mr):** Davdeka/Shutterstock; **p121(b):** The natures scholar/Shutterstock; **p123:** RDVector/Shutterstock; **p126:** Lunov Mykola/Shutterstock; **p127:** Jwensly Dmello/Shutterstock; **p130:** David Aleksandrowicz/Shutterstock; **p131:** Anton Balazh/Shutterstock; **p140:** 3523studio/Shutterstock; **p144:** grass-lifeisgood/Shutterstock; **p145(t):** kentoh/Shutterstock; **p145(b):** Avigator ortuner/Shutterstock.

Artwork by Daniel Limon, Q2A Media, Alan Rogers, Pantek Media, and OKS Prepress.

Every effort has been made to contact copyright holders of material reproduced in this book. Any omissions will be rectified in subsequent printings if notice is given to the publisher.

Contents

Unit 1 Think maths   **5**
Think about how you learn.....................5
Talk to yourself.....................6
How we talk about maths matters.....................7

Unit 2 Number and place value   **8**
Numbers above 9999.....................8
What is a million?.....................9
Represent numbers in different ways.................10
Compare numbers.....................11
Compare and order numbers.....................12
Round numbers.....................13
More rounding.....................14
Number sequences and rules.....................15
Powers of 10.....................16
Count in powers of 10.....................17
Roman numerals.....................18
Work with Roman numerals.....................19

Unit 3 Properties of shapes   **20**
Revisit polygons.....................20
3D shapes and nets.....................21
Match 3D shapes to nets.....................22
Draw 3D shapes.....................23
Classify angles.....................24
Measure angles in degrees.....................25
Draw and measure angles.....................26
Angles in triangles and quadrilaterals.............27
Compare angles.....................28
Angles on a straight line.....................29
Calculate unknown angles.....................30
Shape patterns and sequences.....................31

Unit 4 Addition and subtraction   **32**
Mental calculation.....................32
Count on to add.....................33
Count in steps to subtract.....................34
Find pairs.....................35
Add larger numbers.....................36
Subtract larger numbers.....................37
Inverse operations.....................38
Estimate and give approximate answers.........39
Estimate when adding and subtracting.............40
Multi-step problems.....................41

Unit 5 Decimals and percentages   **42**
Revisit decimal place value.....................42

Thousandths.....................43
Represent decimals in different ways.................44
Compare and order decimals.....................45
Round decimals.....................46
More decimals.....................47
Percentages.....................48
Percentages, decimals and fractions.................49
Find a percentage of an amount.....................50
Equivalent percentages, fractions
and decimals.....................51

Unit 6 Time   **52**
Time zones.....................52
Convert units of time.....................53
Less than a second.....................54
Calculate with times.....................55

Mixed practice 1   **56**

Unit 7 Multiplication and division 1   **58**
What do you already know?.....................58
Multiples.....................59
Sequences of multiples.....................60
Use multiples to solve problems.....................61
Square numbers.....................62
Cube numbers.....................63
Factors.....................64
Find factors of a number.....................65
Common factors and multiples.....................66
Prime numbers.....................67
Multiply by 10, 100 or 1000.....................68
Divide by 10, 100 or 1000.....................69
Make multiplying simpler.....................70
Write a number as a product of its factors.......71
Order of operations.....................72
Work in order.....................73

Unit 8 Measures and money   **74**
Length.....................74
Mass.....................75
Capacity.....................76
Measuring scales.....................77
Read measuring scales.....................78
Metric and imperial equivalents.....................79
Solve measurement problems.....................80
Work with money.....................81

Unit 9 Perimeter and area — 82

Perimeter...82
Composite shapes..83
Area..84
Use a formula to calculate area........................85
More area calculations...................................86
Solve perimeter and area problems....................87

Unit 10 Statistics — 88

Use data to answer questions..........................88
Tables and graphs..89
Tables and diagrams.....................................90
Line graphs...91
More line graphs...92
The mode, median and range...........................93
Frequency tables...94
Frequency tables with groups..........................95
Do your own investigation..............................96
Timetables..97

Unit 11 Fractions — 98

Revisit fractions...98
Equivalent fractions.....................................99
Improper fractions and mixed numbers...............100
Compare and order fractions..........................101
Add and subtract fractions............................102
More adding and subtracting..........................103
Multiply fractions by whole numbers..................104
Multiply with mixed numbers..........................105
Fractions of an amount.................................106
More work with fractions...............................107

Unit 12 Position, direction and movement — 108

Coordinates on a grid...................................108
Shapes on a grid..109
Reflect shapes on a grid................................110
Translate shapes on a grid.............................111
More moving shapes....................................112
Reflections and symmetrical patterns.................113

Mixed practice 2 — 114

Unit 13 Multiplication and division 2 — 116

Written multiplication...................................116
Multiply by 2-digit numbers............................117
Practise multiplying.....................................118
Revisit division..119

Written division..120
Estimate and divide with remainders..................121
Can you divide the remainder?.........................122
More work with remainders............................123
More division..124
Solve division problems.................................125
Mixed problems...126
What is a rate?..127

Unit 14 Work with negative numbers — 128

Positive and negative numbers.........................128
Count on and back through 0...........................129
Count in steps...130
Number sequences......................................131
Temperature changes...................................132
Add and subtract with negative numbers.............133

Unit 15 Calculate with decimals — 134

Pairs that make 1..134
Use facts to add hundredths...........................135
Add and subtract across a whole number.............136
Add and subtract any decimals........................137
Multiply and divide by 10 and 100.....................138
Multiply decimals by whole numbers..................139
More multiplying..140
Mixed decimal problems................................141

Unit 16 Volume and capacity — 142

Volume of 3D objects...................................142
Think about volume.....................................143
Find the volume of containers.........................144
Calculate volume..145
Solve volume problems.................................146

Unit 17 Ratio and proportion — 147

What is a ratio?...147
What is a proportion?...................................148
Work with ratio and proportion........................149
Solve ratio and proportion problems..................150

Unit 18 Probability — 151

How likely is it?...151
The probability scale....................................152
More probability...153
Experiment with probability............................154

Mixed practice 3 — 155

Glossary — 157

Think maths

Think about how you learn

 Think and share

Sharla made this **pattern** using paper squares that measure 4 cm by 4 cm. She has cut some of the squares. She has two squares left over.

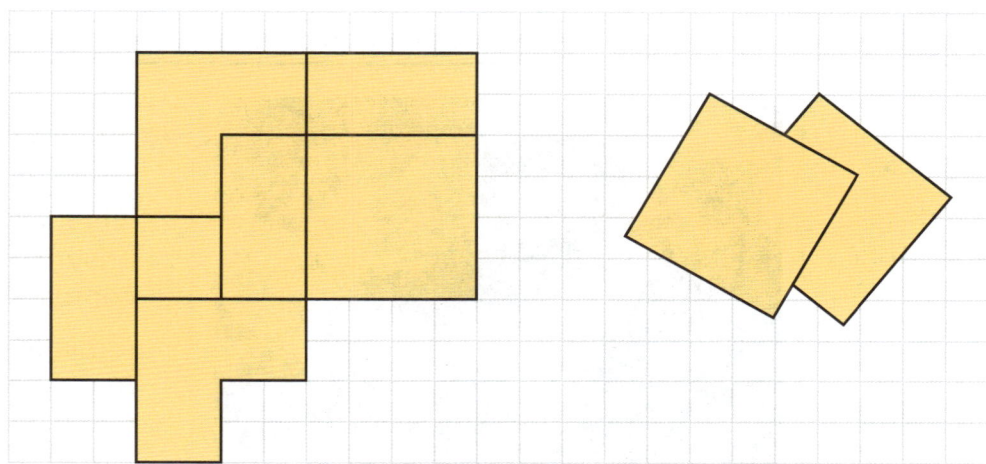

- Work out how many squares Sharla has used in her pattern.
- How did you work this out? Tell your group.
- Did you talk to yourself while you were working it out?
- How can talking to yourself while you work help you?
- How can talking to others while you work help you?
- Can you make the pattern into a square with the two squares left over? How?

1 Copy this shape. How many ways can you find to shade $\frac{1}{8}$ of the shape?

Work on your own to start with and then share your ideas with your group.

2 Think about the problems on this page and the discussions you had. Tell your group two interesting things that you heard or learnt.

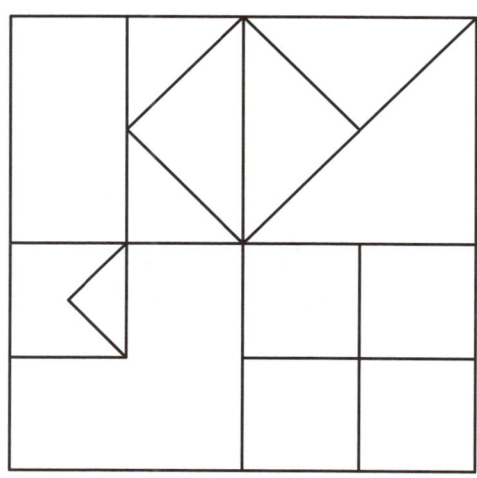

Talk to yourself

Talking to yourself can help your brain process information and help you to think clearly. Here are some of the ways that we can talk to ourselves to help our brains work better.

Planning
- organises our thoughts
- helps us decide what we need to do

Checking in
- helps us to decide if we are on the right track
- can lead to a change in our approach

Making connections
- reminds us about other times when we've done similar things
- links maths to other areas and ideas

Struggle talk
- lets us know when we are unsure or confused
- reminds us to get help and work out what to do

Strategising
- focuses on the steps in the maths we are doing

Encouraging
- helps keep us going
- reminds us that we have good skills and shouldn't give up

Focusing
- reminds us to concentrate and not get distracted

---

1 Zayne arranged the letters of his name in this connecting diagram. Consecutive letters are not connected to each other.

> Consecutive means following one after the other.

a Draw a connecting diagram like this for your name. As you work, notice how you talk to yourself about the task.

b What was challenging about this task? Tell your group what you did.

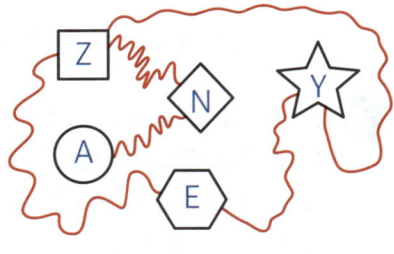

➡ *Workbook page 5*

How we talk about maths matters

SUCCESS

- I can learn anything I want to.
- When I fail, I learn.
- When I'm stuck, I keep going
- I like to challenge my brain.
- My effort and attitude help me succeed.

- What important messages do these posters give?
- Why are these messages important?

When we talk positively about ourselves and our work, we are more likely to learn and succeed.

1 Maths pupils with a growth mindset have certain beliefs that help them to succeed. For example, they believe that anyone can do maths to a high level, if they work hard and have the right mindset.

Work in groups to complete these growth mindset statements.

a
Intelligence can grow
as long as you …

b
You should keep trying when
things are hard because …

c
Challenging problems
help you to …

d
Making mistakes is
important because …

e
The process of learning is
more important than the
result because …

f
Sharing your thinking
helps …

➡ *Workbook page 6*

Number and place value

Numbers above 9999

Think and share

Read and discuss the information about the world's longest cruise ship.

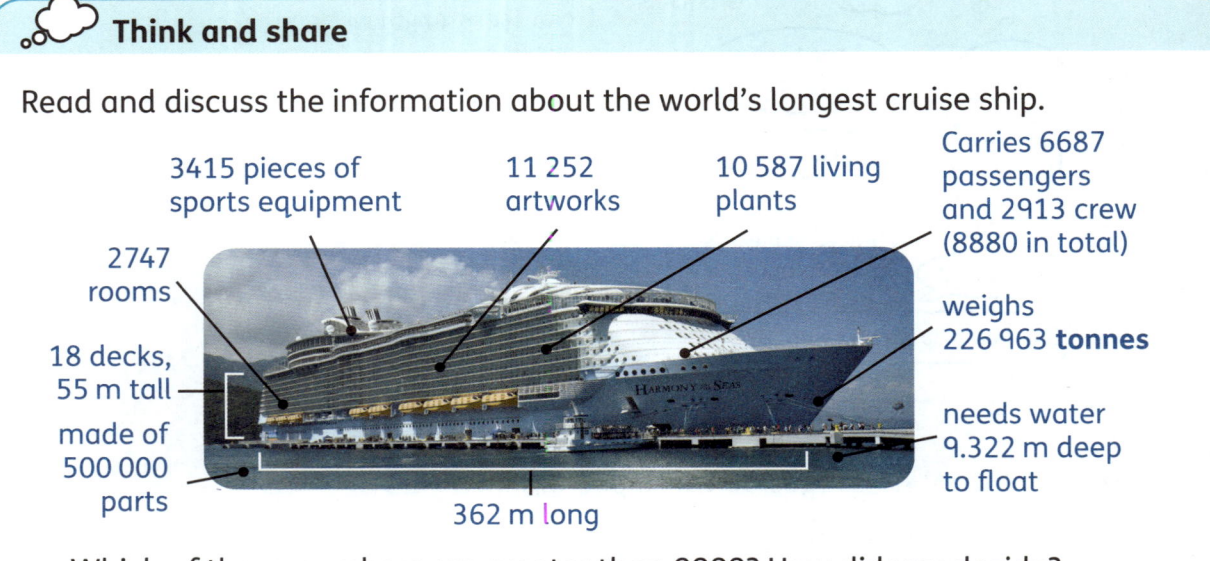

- 3415 pieces of sports equipment
- 11 252 artworks
- 10 587 living plants
- Carries 6687 passengers and 2913 crew (8880 in total)
- 2747 rooms
- 18 decks, 55 m tall
- made of 500 000 parts
- weighs 226 963 **tonnes**
- needs water 9.322 m deep to float
- 362 m long

- Which of these numbers are greater than 9999? How did you decide?

In a place-value table, the place to the left of thousands is called ten thousands. We write 10 000.

The place to the left of ten thousands is called hundred thousands. We write 100 000.

Hundred thousands	Ten thousands	Thousands	Hundreds	Tens	Ones
	5	6	2	8	1
3	2	4	1	0	9

56 281 is fifty-six thousand, two hundred and eighty-one.

324 109 is three hundred and twenty-four thousand, one hundred and nine.

1. Say each number.

   a 59 802   b 45 195   c 120 000   d 499 099   e 129 004

2. Write in numerals:

   a twelve thousand, three hundred and eighty-six

   b thirty-five thousand, eight hundred and fifty

   c two hundred and fifty-one thousand, three hundred and twenty-eight

Workbook page 7

What is a million?

1 + 999 999 = 1 000 000
1 000 000 is one thousand thousands.
The mathematical name for one thousand thousands is one **million**.

Millions	Hundred thousands	Ten thousands	Thousands	Hundreds	Tens	Ones
	9	9	9	9	9	9
1	0	0	0	0	0	0

1 234 543 is one million, two hundred and thirty-four thousand, five hundred and forty-three.

1 Say each number. Write the value of the underlined, red digit in each one.

a 9̲99 567　　b 1̲ 000 000　　c 1 2̲00 000　　d 7 43̲2 546

e 239̲ 876　　f 1 4̲50 000　　g 2̲ 999 875　　h 4̲ 034 213

2 Read these numbers and then write them in order from smallest to greatest.

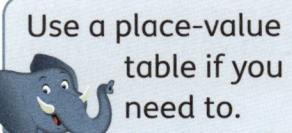
Use a place-value table if you need to.

a 2 108 434　　b 96 103 324　　c 5 443 549

d 5 092 548　　e 8 543 546　　f 5 453 690

💡 Problem solving

3 This meter measures electricity use in kilowatts. It has dials with a pointer that turns. Each dial represents one place in the number.

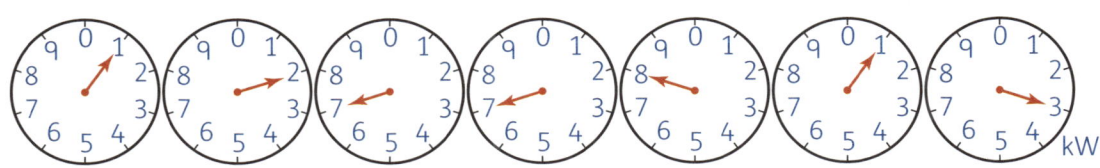

a What is the reading on the dial that shows thousands?

b Why do the dials only go up to 9?

c What is the meter reading of number of kilowatts used?

d What will the meter reading be after another 10 000 kilowatts are used?

➡ *Workbook page 8*

Represent numbers in different ways

Here are some different ways of representing the number 134 589 visually.

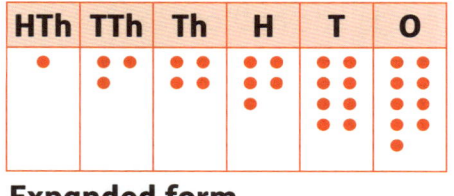

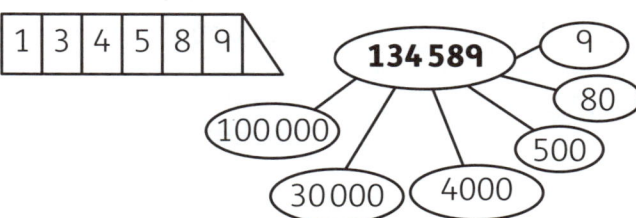

Expanded form
100 000 + 30 000 + 4000 + 500 + 80 + 9 = 134 589

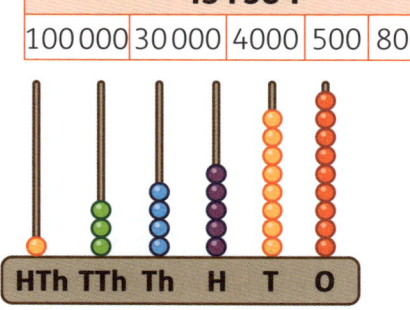

One hundred and thirty-four thousand, five hundred and eighty-nine

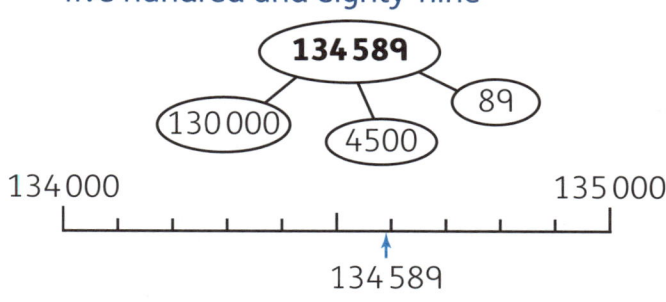

Can you think of any other ways to represent 134 589?

1. Write each of these numbers in expanded form and in words.

   a 12 345   b 91 008   c 124 980   d 403 112

2. What is the value of the digit 5 in each number?

   a 15 702   b 44 520   c 88 959   d 20 005

   e 345 098   f 597 341   g 450 987   h 5 342 123

3. Represent each of these numbers visually. Use a different way to show each one.

   a 12 345   b 980 402   c 234 000   d 192 988

💡 **Problem solving**

4. How many different 6-digit numbers can you make using these cards? Use all six cards in each number.

   | 4 | 5 | 5 | 6 | 2 | 0 |

   Make a set of cards to model the numbers.

Compare numbers

Which representation shows the greatest number?

A

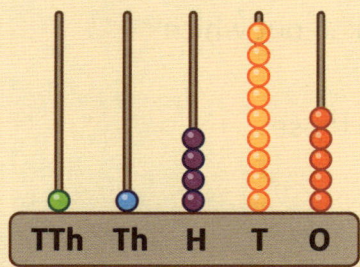

TTh Th H T O

B

12 000	90

C

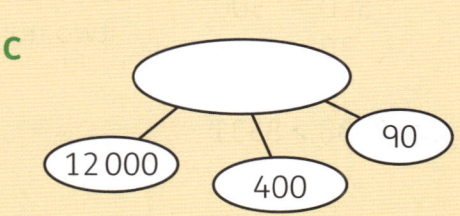

12 000   400   90

D

Twelve thousand, four hundred and ninety-three

How did you decide?

You can use **place value** to help you compare numbers.

Which number is greater, 8946 or 8765?

8 9 4 6   Write the numbers so that the same places are below each other.
8 7 6 5   Compare digits, starting at the left. Stop at the first digit that
is different.
9 hundreds are more than 7 hundreds, so 8946 is greater
than 8765.

Which number is greater, 13 456 or 13 465?

1 3 4 5 6
1 3 4 6 5   6 tens are more than 5 tens, so 13 465 is greater than 13 456.

---

1 Copy the number statements. Write < or > to compare each pair of numbers.

a 4012 ☐ 5000

b 6825 ☐ 5149

c 9060 ☐ 9600

d 12 345 ☐ 12 400

e 13 400 ☐ 14 300

f 19 350 ☐ 19 349

g 234 453 ☐ 234 543

h 908 000 ☐ 908 700

i 124 786 ☐ 124 876

> Remember:
> < means less than
> and > means
> greater
> than.

Compare and order numbers

Bhavna compared two numbers using place value.

8538   8531
8000 = 8000
500 = 500
30 =   30
8 >   1
So, 8538 > 8531

Explain how Bhavna worked out which number was greater.

How would you compare these two numbers? Why?

1 Compare the numbers. Fill in < or >.

a 2484 ☐ 4284

b 1000 ☐ 10 000

c 8402 ☐ 8765

d 20 545 ☐ 20 554

e 46 543 ☐ 465 430

f 124 876 ☐ 124 788

2 Rewrite each set of numbers in order from smallest to greatest.

a 13 168   11 927   12 635

b 26 761   26 716   26 671

c 215 430   214 530   213 450

d 129 642   127 849   126 301

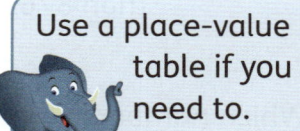
Use a place-value table if you need to.

💡 **Problem solving**

3 Find three different ways to add four beads to each abacus so that the number statements are still correct.

a

337 329 >

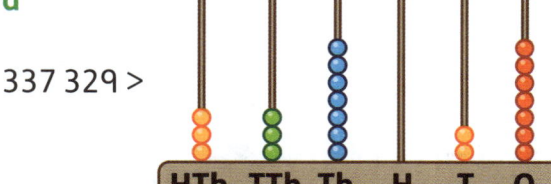

b

934 432 >

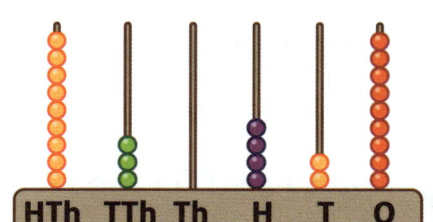

Round numbers

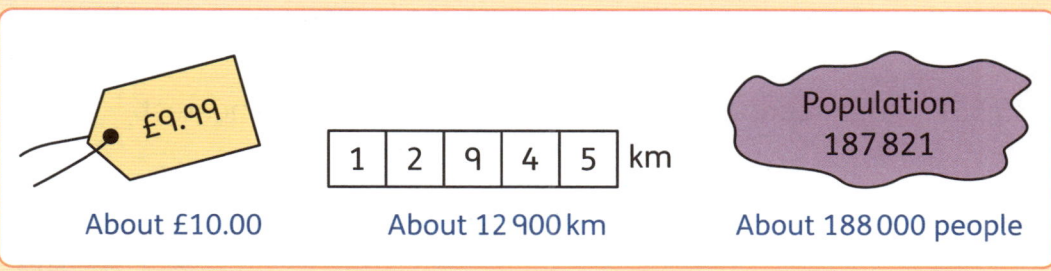

£9.99 — About £10.00

| 1 | 2 | 9 | 4 | 5 | km — About 12 900 km

Population 187 821 — About 188 000 people

You can **round** numbers to the nearest 10, 100 or 1000 using place value and these rules:

- Find the digit with the place value you are rounding to.
- Look at the digit to the right of the one you are rounding to.
- If the digit to the right is 5 or greater, round your digit up.
- If the digit to the right is less than 5, your digit does not change.
- Write 0 as a place holder for every digit to the right of the one you rounded to.

1 Round each number to the nearest 10 and to the nearest 100.

| a | 387 | b | 980 | c | 7255 | d | 3999 |
| e | 12 345 | f | 34 543 | g | 231 435 | h | 119 899 |

2 a Look at the box. List all the numbers that will give 9000 when they are rounded to the nearest 1000.

| 8098 | 8890 | 8499 | 9023 |
| 9500 | 8500 | 10 023 | 9499 |

b Round all the other numbers to the nearest 1000.

3 Work with a partner. Decide whether each statement is true or false.

a 6731 rounds to 7000.　　　　b 18 458 rounds to 18 500.

c 13 987 rounds to 15 000.　　　d 139 800 rounds to 14 000.

Problem solving

4 A whole number is rounded to the nearest 100 to get 17 400.

What are the smallest and greatest numbers that could have been rounded?

➡ *Workbook page 9*

More rounding

You can round numbers to any place using the rules for rounding.

Round 135 567 to the nearest 10 000.

Ten thousands place
↓

135 567
↑

The digit to the right is 5, so round up the 3.

140 000   Write 0 in all places to the right.

Round 1 421 987 to the nearest 100 000.

Hundred thousands place
↓

1 421 987
↑

The digit to the right is 2, so the 4 does not change.

1 400 000   Write 0 in all places to the right.

1   Look at the box.

  a   List all the numbers that round down to 40 000 to the nearest 10 000

  b   List all the numbers that round up to 50 000 to the nearest 10 000.

41 709	47 532	45 000
42 659	48 518	43 598
46 729	44 999	

2   Round each number to the nearest 10 000.

  a   123 456   **b**   14 876   **c**   19 045   **d**   12 999

3   Round each population figure to the nearest 100 000.

Bulgaria	Lebanon	Singapore	Finland
6 948 455	6 825 445	5 850 445	5 518 087
New Zealand	**Kuwait**	**Mongolia**	**Bahrain**
4 822 233	4 270 571	3 278 290	1 701 575

Problem solving

4   Match pairs of numbers that will give a total of 1 000 000 when they are rounded to the nearest 100 000.

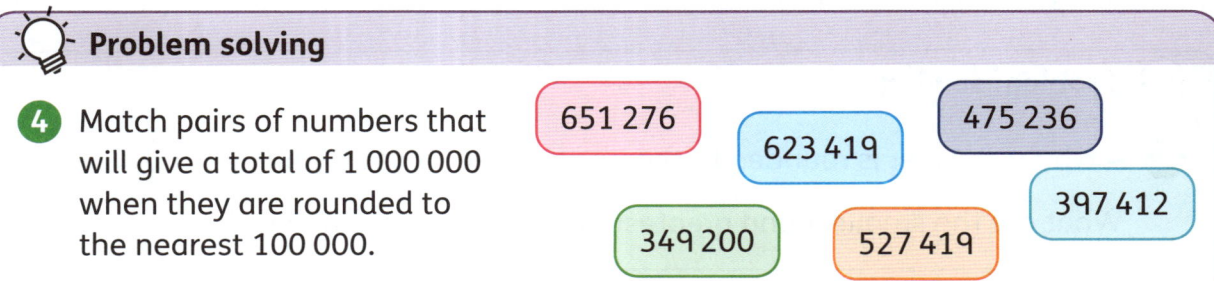

651 276   623 419   475 236   397 412   349 200   527 419

➡ *Workbook page 10*

Number sequences and rules

A number **sequence** (or pattern) is an ordered set of numbers in **ascending** or **descending** order. Each number in a sequence is called a **term**.

200, 400, 600, 800, ...

This sequence is in ascending order. Each term is 200 greater than the one before it.

The next three terms are 1000, 1200 and 1400.

1000, 950, 900, 850, ...

This sequence is in descending order. Each term is 50 less than the one before it.

The next three terms are 800, 750 and 700.

We can describe number sequences using the first term and the **term-to-term rule**. For the sequence 200, 400, 600, 800, ... the first term is 200, the term-to-term rule is add 200.

1 Work out the term-to-term rule for each sequence. Write the rule and the next three terms.

a 30 142, 30 144, 30 146, 30 148, ... b 700, 1400, 2100, ...

c 10, 100, 1000, 10 000, ... d 6000, 12 000, 24 000, 48 000, ...

2 Write the first five terms of each sequence.

a First term is 1 000 000. Count back 10 000 to get to the next term.

b First term is 450 000. Add 10 000 to get to the next term.

c First term 37. Term-to-term rule: subtract 3.

d First term 28. Term-to-term rule: add 5.

Problem solving

3 Cicadas are insects that hatch only every 13 or 17 years. In 2021, Brood X hatched for the first time in 17 years.

a List the next five years in which Brood X cicadas will hatch.

b A brood of cicadas that hatch every 13 years also hatched in 2021. When will the two broods next hatch at the same time?

➡ Workbook page 11

Powers of 10

This sequence of numbers shows the first six **powers of 10**.

| 10 | 100 | 1000 | 10 000 | 100 000 | 1 000 000 |

Look at the sequence. What do you think powers of 10 means?
What will the next number in the sequence be? Why?

1 For each number in the box, write the number that is:

    a 10 more   **b** 100 less

    c 10 000 more **d** 1000 less

| 23 453 | 19 532 | 31 000 |
| 530 345 | 454 900 | |

2 These number sequences have been mixed up. Rewrite them in ascending order. Then write the term-to-term rule for each sequence.

    a 26 400   24 400   23 400   25 400   27 400

    b 162 402   152 402   132 402   172 402   142 402

    c 894 320   904 320   884 320   864 320   874 320

3 Read each rule. Write the first five numbers in each sequence.

    a Count back in 10 000s. End at 13 450.

    b Start at 1 000 000. Count back in 100 000s.

Problem solving

4 Find your way through this grid by counting back in powers of 10.
You can move up, down, left or right. List the powers of 10 that you count back each time.

START →

432 960	432 965	432 449	332 450	318 630	306 730
431 960	431 950	431 920	309 604	308 630	307 630
431 940	421 950	421 955	309 640	308 640	306 630
419 950	420 950	319 740	309 740	206 530	206 630
319 950	319 850	319 750	319 700	205 530	205 330
319 960	319 650	319 550	319 500	204 530	104 530

➡ *Workbook page 12*

Count in powers of 10

1 The graph shows how many games a new business expects to sell in the first 7 months.

   a How many games do they expect to sell in the first month?

   b How many games do they expect to sell each month after that?

   c If this pattern continues, in which month will they first sell more than 100 000 games?

   d From the start of month 2, their actual sales are 10 times more than they expected. Work out how many games they sold in months 2, 3, 4, 5, 6 and 7.

2 Write the number that is:

   a 10 000 more than 143 000

   b 1 million more than 34 000

   c 100 000 more than 5689

   d 1000 less than 234 567

   e 10 less than 129 000

   f 10 000 less than 892 000

3 Write the operations, using powers of 10, that will get you from the first number to the second. The first one has been done as an example.

   a 234 543 → 244 643   + 10 000 + 100

   b 465 800 → 355 800

   c 999 900 → 889 800

   d 23 895 → 1 123 895

   e 9875 → 109 965

4 Ranjeet has written these number sequences. There is at least one mistake in each one. Work with a partner to find the mistakes and say what you think Ranjeet has done wrong in each sequence.

a	159 999	14 999	129 999	119 999	19 999
b	600 000	590 000	580 000	57 000	580 000	590 000
c	499 123	399 123	299 123	199 123	100 123	123
d	23 300	24 300	25 300	23 400	27 300
e	976 524	876 524	776 524	677 524	576 524

Roman numerals

What do you think the sign on this building tells us? Can you read the number?

The number on the building is a year written in Roman numerals.

Roman numerals are written using combinations of these seven letters.

I	V	X	L	C	D	M
1	5	10	50	100	500	1000

The Roman number system does not use place value and it has no zero (0). There are rules to make different numbers.

Rule	Example
If the letters are the same, and next to each other, you add them.	XXX = 10 + 10 + 10 = 30 CC = 100 + 100 = 200
Letters cannot be written more than three times in a row.	XX = 20 CCC = 300 14 = XIV not XIIII
If letters of a lower value are to the right of letters of a higher value, you add them.	LXX = 50 + 10 + 10 = 70 CCL = 100 + 100 + 50 = 250
If letters of a lower value are to the left of letters of a higher value, you subtract them.	XL = 50 − 10 = 40 XC = 100 − 10 = 90
If a letter of lower value is between two letters of higher value, subtract it from the one on the right.	XIV = 10 + (5 − 1) = 14 XXIX = 10 + 10 + (10 − 1) = 29

1 Write these Roman numerals as ordinary numbers.

a IV   b IX   c XL

d LX   e XC   f CI

2 Write the missing number in each sequence using Roman numerals.

a ☐ 100 150 200 250   b 120 160 ☐ 240 280

c 201 301 401 ☐ 601   d 1200 1100 1000 ☐ 800

e 25 50 ☐ 100 125   f 99 199 299 399 ☐

➡ *Workbook page 13*

Work with Roman numerals

 Problem solving

1 These distance markers are from Roman roads. They show how many kilometres each place is from Rome.

PISA
CCLXIV

TURIN
DXXIV

OHRID
DCC

MILAN
CDLXXVII

TUNIS
DXCIX

GENOA CDII

NAPLES
CLXXXVIII

TRIESTE
CDXXX

LJUBLJANA
CDLXXXIX

BIZERTE
DLX

a Work out each distance.

b List the places in order of distance from Rome, starting with the shortest distance.

2 At the end of a TV programme or movie, you sometimes see the year written in Roman numerals.

a What year is shown on each of these screens?

FILMED ENTIRELY
IN INDIA

© MMXX

a mad dragon
production
© MMXVIII

DOWN UNDER
FILMS

© MMXXIV

b Write your year of birth in Roman numerals.

Properties of shapes

Revisit polygons

  Think and share

A **polygon** is a closed shape with straight sides that meet at **vertices** (corners).

Polygons with all sides equal in length and all **angles** equal in size are called regular polygons.

Which of these polygons are regular? Tell your partner how you decided.

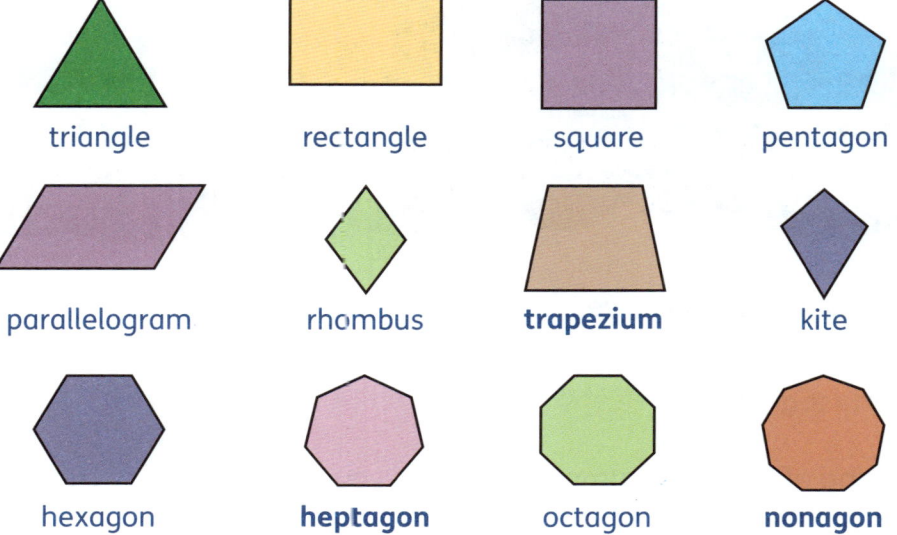

| triangle | rectangle | square | pentagon |

| parallelogram | rhombus | **trapezium** | kite |

| hexagon | **heptagon** | octagon | **nonagon** |

Take turns to test each other's knowledge of polygons. Say the name of a polygon and ask your partner to describe its properties. Or describe the properties and ask your partner to name the polygon.

1 Draw these polygons. Write the name of each shape.

a   A regular quadrilateral

b   An irregular quadrilateral

c   An irregular triangle

d   An irregular pentagon

e   A quadrilateral made from two regular triangles

f   A quadrilateral made from two regular quadrilaterals

g   A regular quadrilateral made from two rectangles

➤ *Workbook page 14*

3D shapes and nets

A 3D shape or box can be made from a 2D shape called a **net**. When you cut open a box and flatten it, you can see the net of the box.

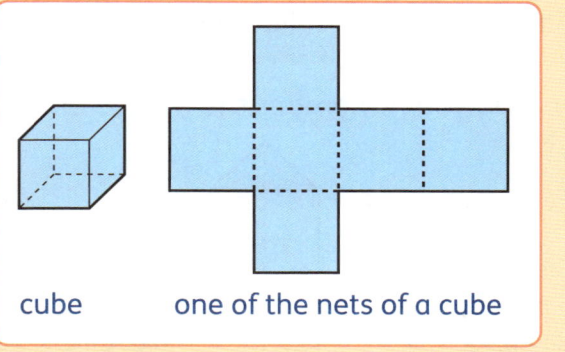

cube   one of the nets of a cube

1 Which of these nets can be used to make a cube? If you are not sure, draw each net on squared paper, cut it out and fold it up to see if it works.

a
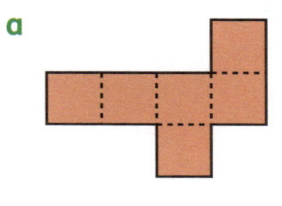

b
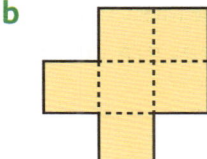

c
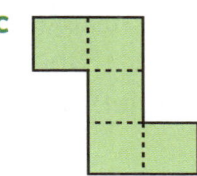

d
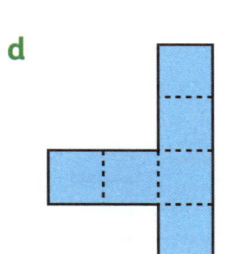

e
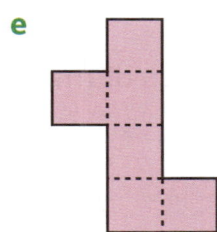

f
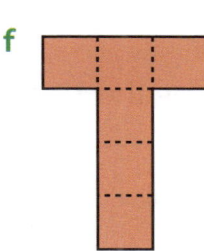

2 Compare each of these nets with the net of a cube. Answer the questions about each net.

A
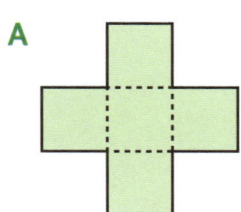

B
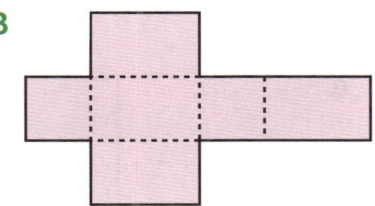

a   How is it similar?

b   How is it different?

c   What 3D shape will you get if you fold up this net?

➡ *Workbook page 15*

Match 3D shapes to nets

Here are some solid 3D shapes.

Do you remember the names of these 3D shapes?

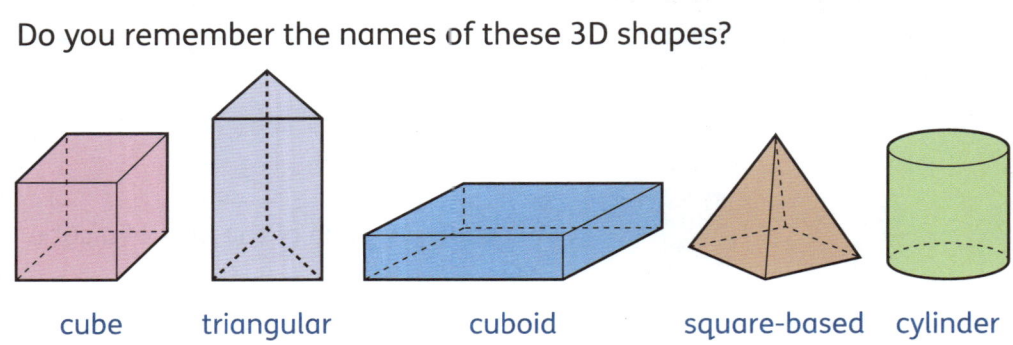

cube   triangular prism   cuboid   square-based pyramid   cylinder

Remember:
the flat surfaces of a solid are called faces
the places where two faces meet are called edges
the corners are called vertices.

1 How many faces, vertices and edges does each 3D shape above have?

2 Name the 3D shape made by each of these nets.

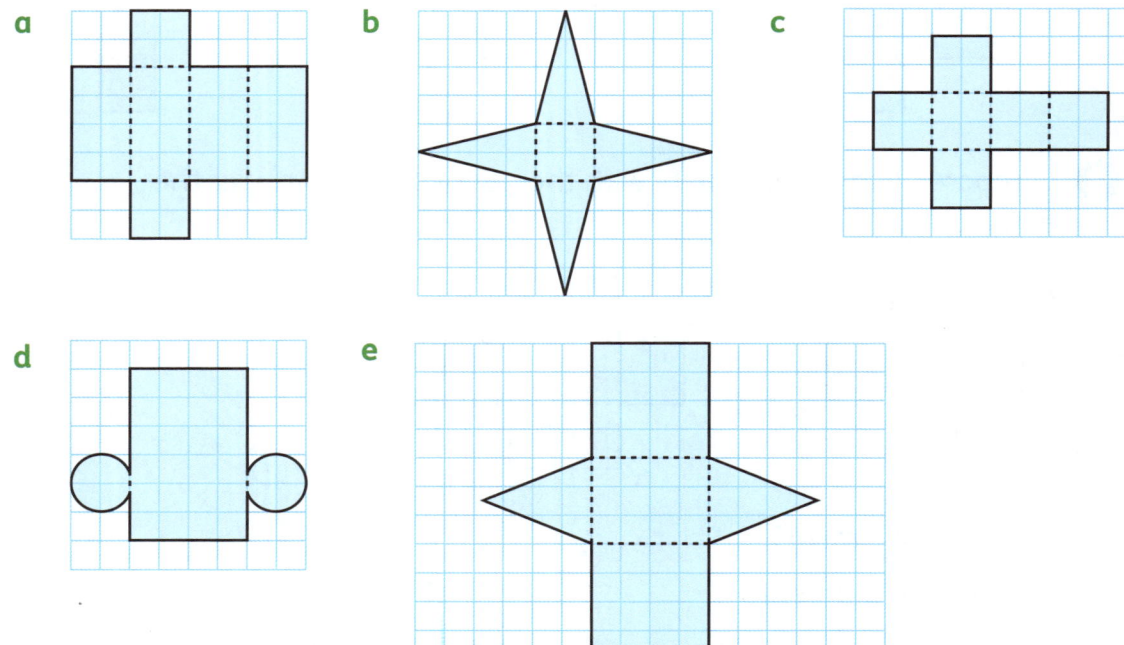

a   b   c   d   e

3 Explain how you decided which 3D shape each net would make.

Draw 3D shapes

Work on squared paper to make it easier to keep lines the same length.

There are different ways to draw 3D shapes.

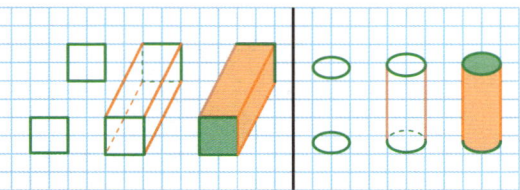

Draw the end faces and join the vertices. Show edges you cannot see with dotted lines. Shading the drawing makes it look more realistic.

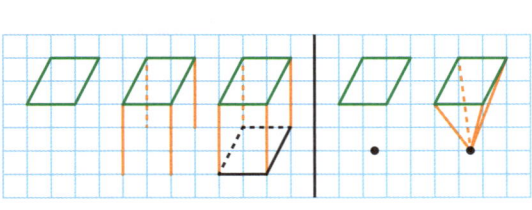

Draw two sets of parallel lines to represent one rectangular face. Draw lines of equal length from each vertex. Draw the base the same shape as the top. You can also draw lines from the vertices to meet at a point.

Square grids show the shape viewed from one face.

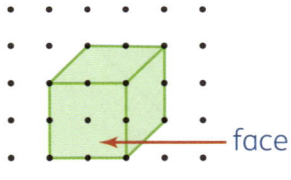

face

Triangular grids show the shape viewed from one edge.

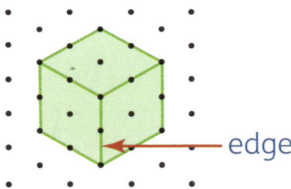

edge

1. Work with a partner. Talk about each method and then try to draw a 3D shape using it.

2. The pyramid in the example above points downwards.

   a Draw a pyramid pointing upwards.

   b Can you draw a pyramid on its side using this method?

3. One of the end faces of three different solids is shown below. Draw what the solid could look like. Use the method that you find easiest.

a 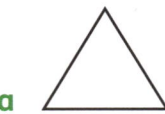 b 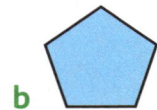 c

➡ *Workbook pages 16–17*

Classify angles

An angle is a measure of turn. The size of an angle is measured in degrees. The sign for degree is °.
Angles are classified according to their size.

A right angle is equal to 90°.

A full turn is 360°. This is the same as four right angles.

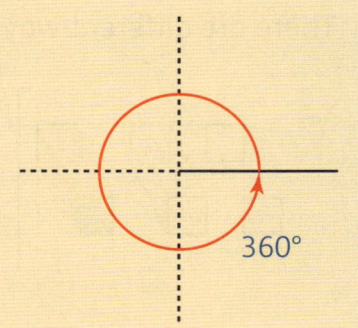

360°

 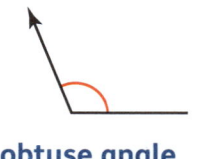 

right angle
90°

acute angle
less than 90°

obtuse angle
greater than 90°

reflex angle
greater than 180°
but less than 360°

1 Name each polygon. Then list how many acute, obtuse and right angles each shape contains.

a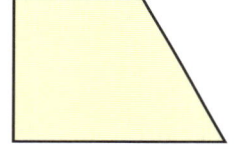

b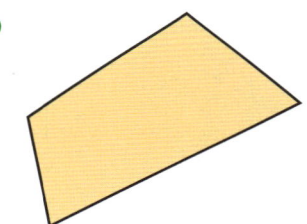

c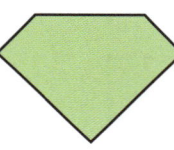

d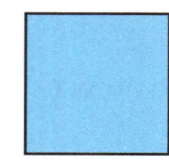

e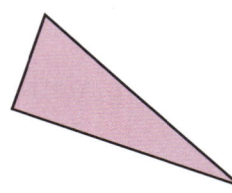

f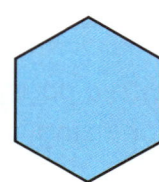

2 Classify each angle as acute, right, obtuse or reflex.

a

b

c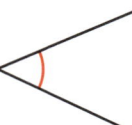

d

e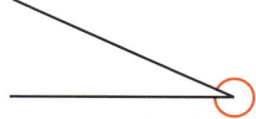

f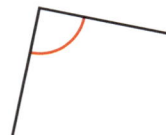

➡ *Workbook page 18*

Measure angles in degrees

A protractor is an instrument for measuring angles.

There are ten divisions between the 10° markers on a protractor. They are used to measure angles more accurately. Each fifth division is made slightly longer to help you read the measurement to the nearest 5°.

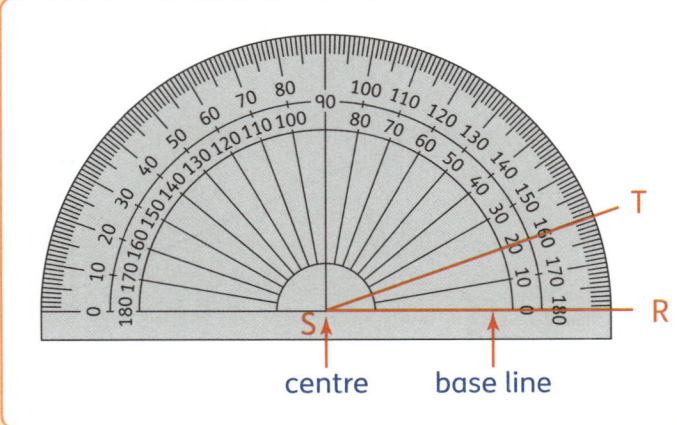

We can write the angle joining line RS to line ST as angle RST.
Angle RST is about 20°.

centre   base line

1. Estimate the size of these angles to the nearest 5°. Then measure them. Write your estimate and measurement in a table.

a

b

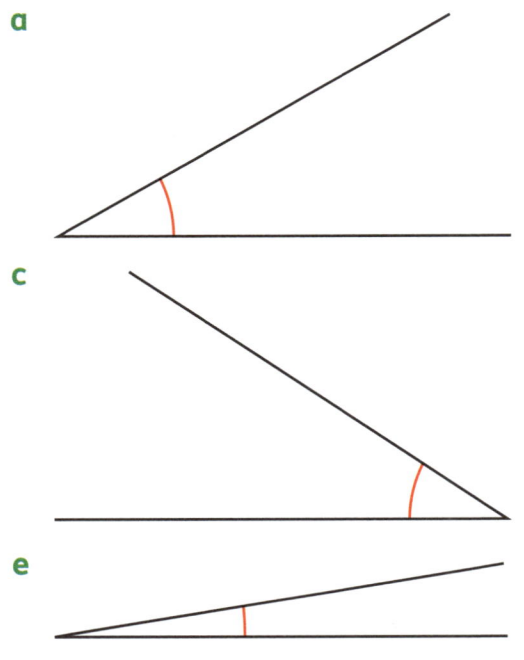

c

d

e

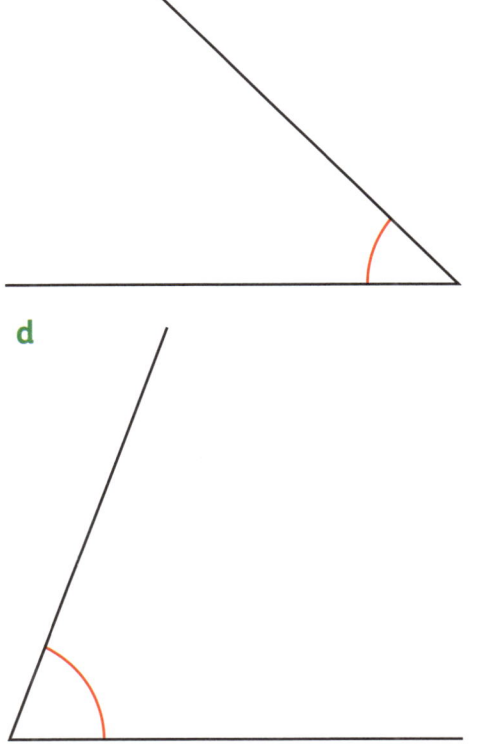

Draw and measure angles

1 Measure these angles and lines carefully. Then draw them accurately.

a

b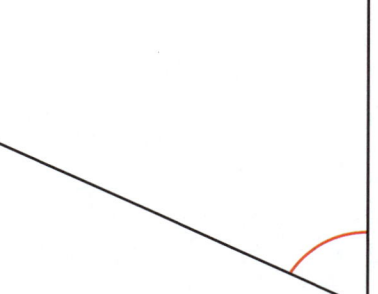

c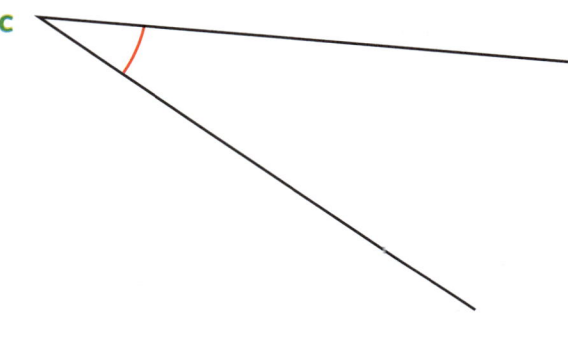

d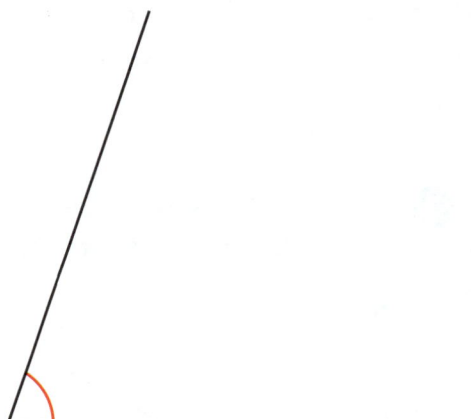

2 Measure each angle. Draw an angle that is 15° greater than each one. Write what type of angle you have drawn.

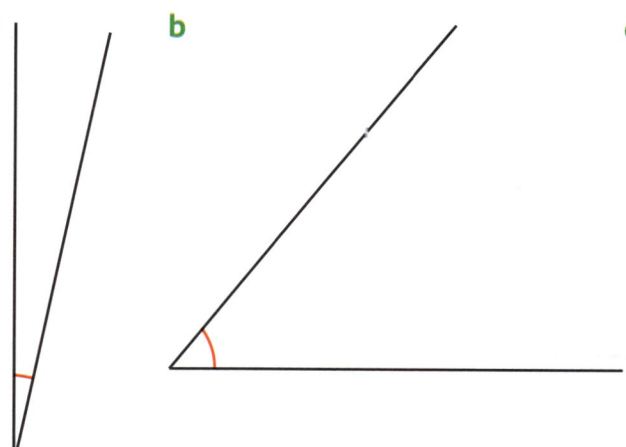

Angles in triangles and quadrilaterals

Triangles are classified according to the lengths of their sides and the sizes of their angles.

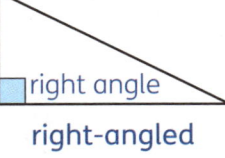

no equal sides

right angle

**right-angled
scalene**

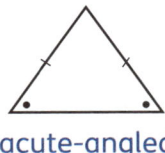

two equal sides and two equal angles

**acute-angled
isosceles**

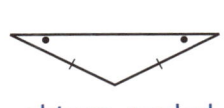

**obtuse-angled
isosceles**

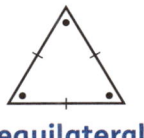

all sides and
all angles equal

equilateral

Quadrilaterals are named according to the properties of their sides and angles.

square

rectangle

rhombus

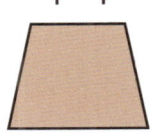

trapezium

parallelogram

kite

1   **a**   Measure and record the length of the sides and the angles of each triangle.

    b   Classify each triangle based on your measurements.

    c   Find the sum of the angles of each triangle. What do you notice?

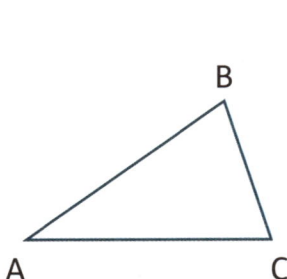

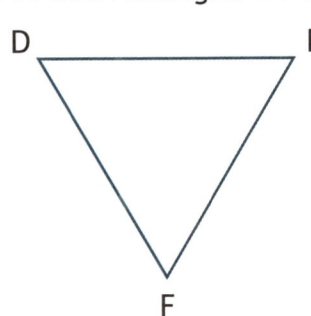

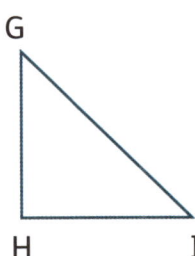

2   Work with a partner to answer these questions about quadrilaterals.

    a   Which quadrilaterals contain only right angles?

    b   Is it possible to have a quadrilateral with two right angles and two angles that are not right angles? Explain why or why not.

    c   Which quadrilaterals have their diagonally opposite angles equal in size?

    d   Which quadrilaterals have four angles that are equal in size?

    e   What is the sum of the angles of any quadrilateral?

Compare angles

1 Which statements about these angles are true?

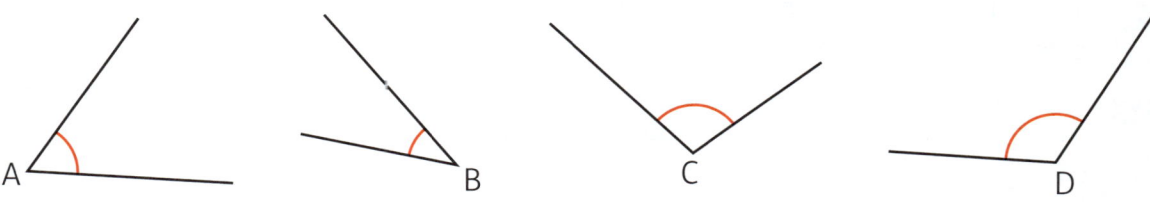

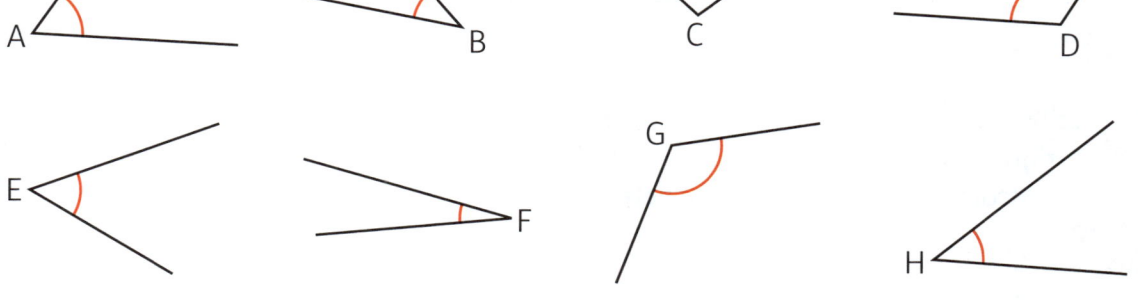

a G is the largest angle.   **b** F is the smallest angle.

c D is a right angle.   **d** A < C

e D > G   **f** F = B

2 Redraw angles A, E, G and H in order, from largest to smallest. Write the measurement next to each angle.

3 Read each statement. Do you agree or not? Give reasons.

a X, Y and Z are different types of angle.

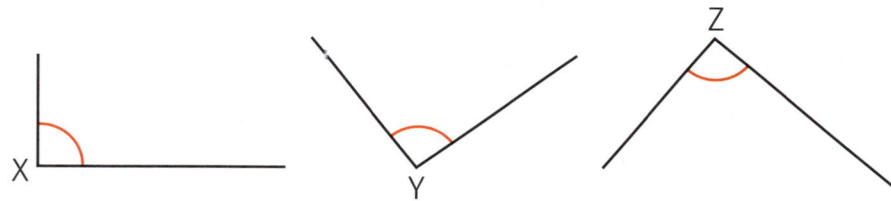

b Angle m is smaller than angle n.   **c** Angle q is the same size as angle p.

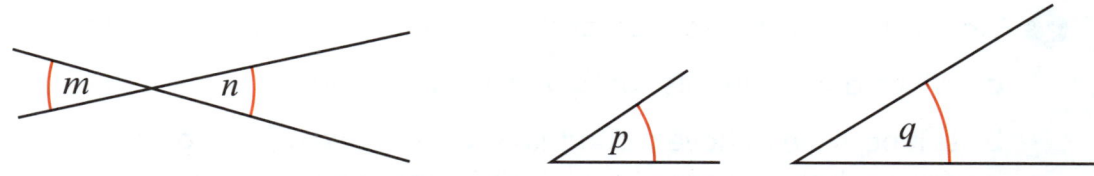

Angles on a straight line

1 Draw a straight line AB.

a Now draw a line CD perpendicular to your straight line AB.

You can use a set square or protractor.

b What are the sizes of the angles CDA and CDB?

c Work out the sum of angles CDA + CDB.

d Complete the rule about angles on a straight line: Angles on a straight line add up to ☐ degrees.

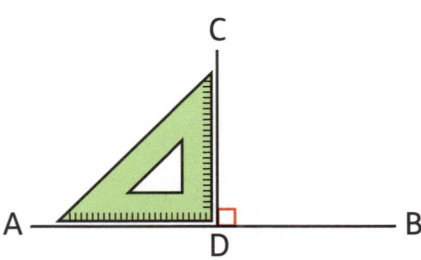

2 Test your rule by drawing some straight lines with a ruler and measuring them using a protractor.

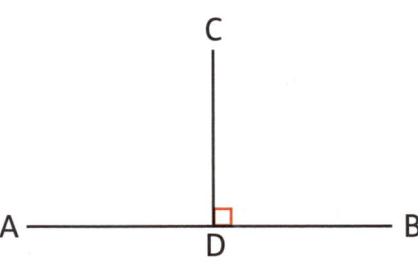

3 Are these angles labelled correctly? Without using a protractor, do a calculation and say yes or no.

a

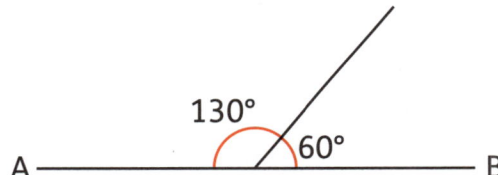

b

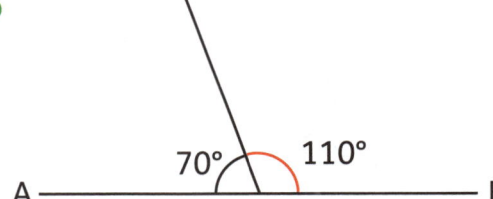

c

d

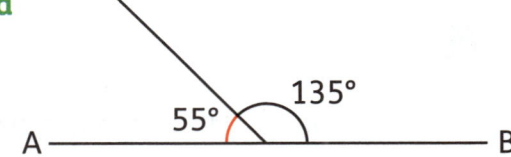

e

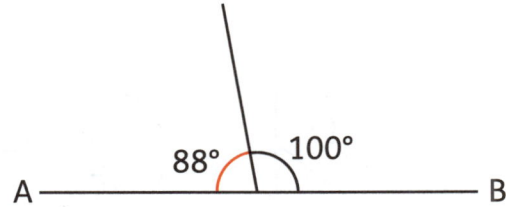

f

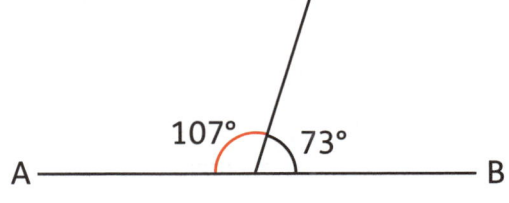

➡ Workbook page 19

Calculate unknown angles

Remember: a full turn is 360°; a half turn, or straight line, is 180°.

What is the size of angle x?

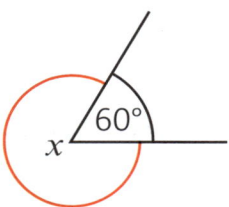

A full turn is 360°.
The known angle is a turn of 60°.
$x + 60° = 360°$
$360° − 60° = 300°$
So, $x = 300°$

What is the size of angle y?

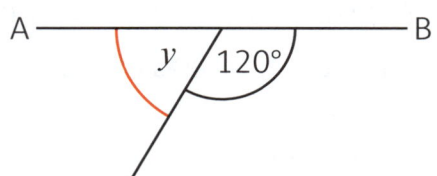

A straight line is 180°.
The known angle is 120°.
$120° + y = 180°$
$180° − 120° = 60°$
So, $y = 60°$

---

1. Work out the size of each angle labelled with a letter.

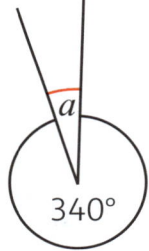
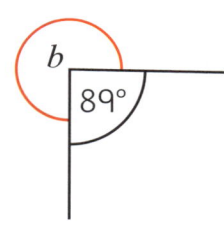
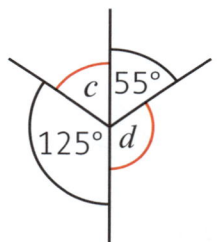
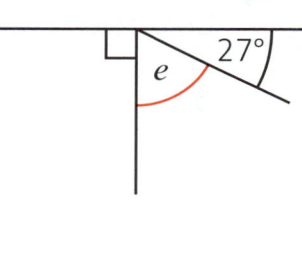

Problem solving

2. Milo is building a shelf for his TV. The shelf is at 90° to the wall. The support wire meets the shelf at an angle of 28°. What are the sizes of angles a and b?

3. Work out the size of the angles marked with letters in each of these diagrams.

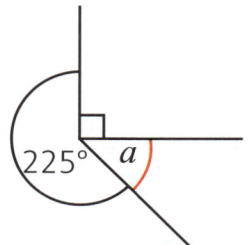
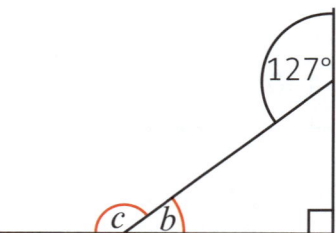
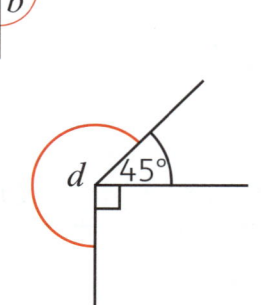

Shape patterns and sequences

Look at this shape pattern.

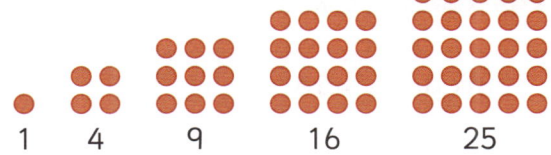

1   4   9   16   25

What do we call these numbers?
What is the next number in the pattern? Why?
What does the tenth shape in the pattern look like? How do you know this?

The pattern above shows the sequence of **square numbers**.

The two patterns below show a different sequence of numbers.

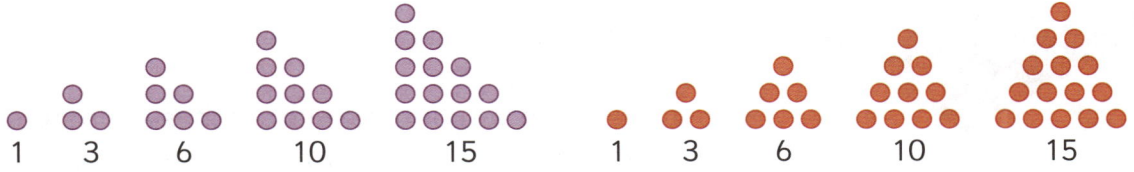

1   3   6   10   15   1   3   6   10   15

Why do you think these are called triangular numbers?

1 The first five triangular numbers are 1, 3, 6, 10, 15.

   a Look at the sequence. Work out the differences between consecutive terms.

   b What is the next term?

   c Follow this pattern to write all the triangular numbers up to 100.

2 Nasif is trying to work out a rule for finding triangular numbers. He remembers that a triangle is half a rectangle, and he draws these shapes.

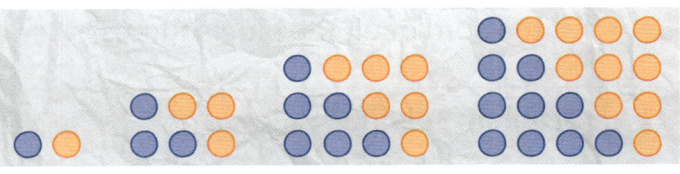

   a Multiply the number of dots in the length by the number in the width of each rectangle. Write this sequence of numbers.

   b Compare your sequence with the sequence of triangular numbers. What do you notice?

   c Can you find a way to use this information to work out the 9th triangular number without drawing any diagrams? Share your ideas with your group.

Addition and subtraction

Mental calculation

💭 Think and share

Look at this number cube.

These jottings show how Isaac worked out the sum of numbers in the orange squares on each face.

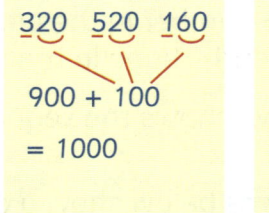

110 120 620

800 + 50

= 850

320 520 160

900 + 100

= 1000

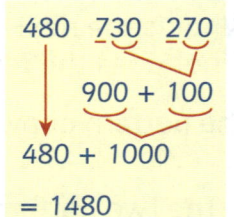

480 730 270

900 + 100

480 + 1000

= 1480

Explain what Isaac did to make it easier to add the numbers mentally.

How else could you find the sum of these numbers? What strategies do you use to add or subtract mentally? Share your ideas.

1 Use the numbers on the cube to do the following calculations mentally. Use a calculator to check your answers.

   a What is the sum of the numbers in the blue squares on each face?

   b Find two purple squares with a difference of 410.

   c Which three orange squares (one from each face) add up to 750?

   d What is the difference between the sum of all the orange squares and the sum of all the blue squares?

💡 Problem solving

2 You can put a + or − sign between some of the digits to make this number sentence true.

1   2   3   4   5   6   7   8 = 90

Here is one way: 1 + 23 + 45 + 6 + 7 + 8 = 90

Find at least one other way to make 90.

➡ *Workbook page 20*

Count on to add

Read through these examples carefully. They show you how to use a blank number line to count on in steps of different sizes.

What is the sum of 1359 + 2147?

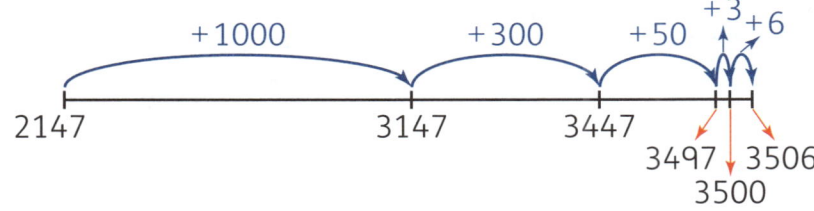

1359 + 2147 = 3506

Start with the greatest number when you count up to add.

What is the sum of 12 938 + 23 849?

12 938 + 23 849 = 36 787

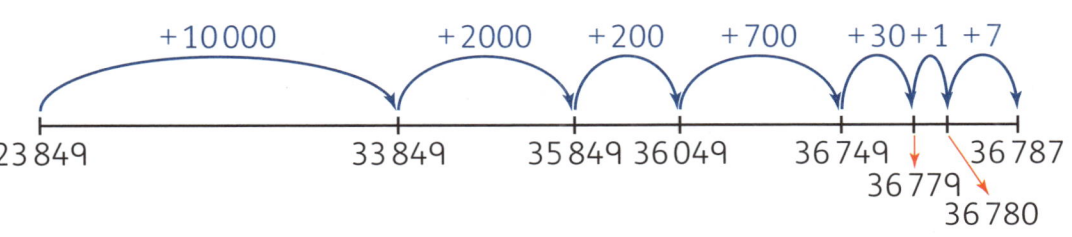

1 Add. Show your working.

a 1234 + 1142

b 1345 + 2366

c 1276 + 3129

d 12 367 + 2876

e 21 987 + 1328

f 12 789 + 4762

2 Here is information about a train company over three days.

	Number of passengers	Income from ticket sales (£)	Distance travelled by all trains (km)
Day 1	10 098	353 430	4550
Day 2	8464	330 096	4098
Day 3	10 103	404 120	4675

a What is the total distance travelled over the three days?

b How many passengers were there in total on days 1 and 3?

c How much more ticket income was there on day 3 than day 2?

d Make up three more addition or subtraction problems using the information in the table. Show how you would solve each one.

Count in steps to subtract

Read through these examples carefully. They show you how to use a blank number line to count back in steps of different sizes.

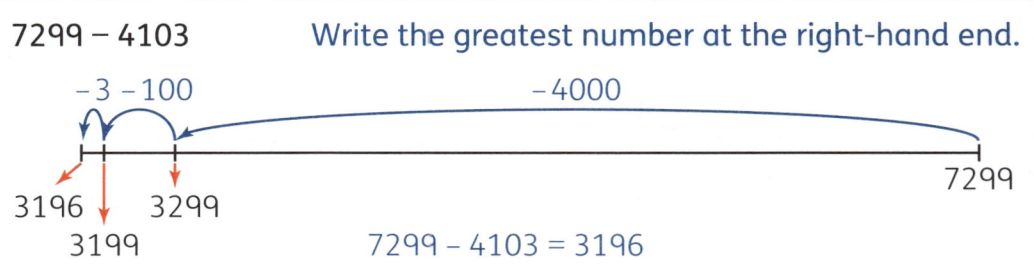

7299 − 4103   Write the greatest number at the right-hand end.

−3 −100 −4000

3196   3299
3199   7299

7299 − 4103 = 3196

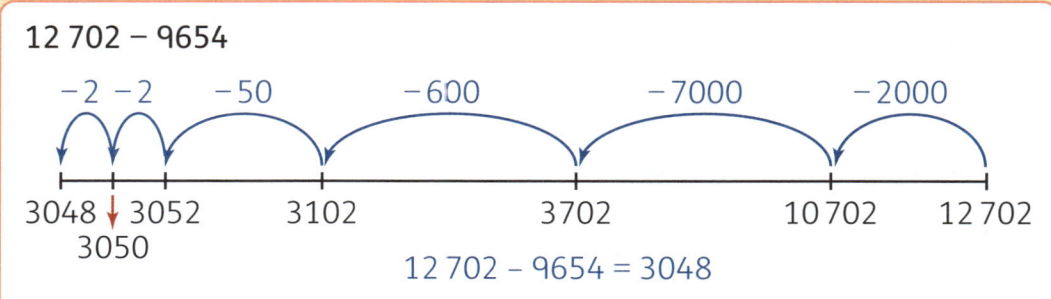

12 702 − 9654

−2 −2   −50   −600   −7000   −2000

3048   3052   3102   3702   10 702   12 702
3050

12 702 − 9654 = 3048

How can you count on to subtract? Give an example to show how.

1 Subtract. Show your working.

   a 1234 − 1004   **b** 3267 − 1034   **c** 3487 − 1876

   d 14 879 − 9456   **e** 21 678 − 11 066   **f** 29 987 − 18 999

💡 Problem solving

2 Here are the **areas** of three island nations:

Iceland
103 000 km²

Ireland
70 273 km²

Sri Lanka
65 610 km²

   a What is the difference in size between the largest and smallest islands?

   b How much bigger is Ireland than Sri Lanka?

   c Ella says Iceland is 30 273 km² bigger than Ireland. Is she correct?

➡ *Workbook page 21*

Find pairs

In this grid, some pairs of numbers have a sum of 10 000, 20 000 or 50 000.

For example, 4620 + 5380 = 10 000.

4620	8560	38 360	2620	6580
8190	11 440	9590	12 540	11 640
1810	10 710	26 460	40 410	9290
7380	5380	37 460	13 420	23 540

1 Find the pairs that total 10 000, 20 000 or 50 000. Use a calculator to check your answers.

2 Work with a partner. Make up your own grid where pairs of numbers total 30 000, 60 000 or 75 000. Swap the grid with another group and ask them to find the pairs.

3 From each set of numbers, find the pair that adds up to a **multiple** of 1000.

a
| 12 150 | 30 570 |
| 30 000 | 37 850 |

b
| 20 800 | 10 990 |
| 30 210 | 49 010 |

c
| 30 840 | 30 890 |
| 42 760 | 57 240 |

d
| 20 000 | 14 870 |
| 40 000 | 28 540 |

e
| 11 000 | 48 650 |
| 28 470 | 19 000 |

4 What patterns did you notice in question 3? Tell your group how these helped you to work out the pairs.

💡 **Problem solving**

There is more than one possible answer.

5 Which containers could you combine to make 1 litre?

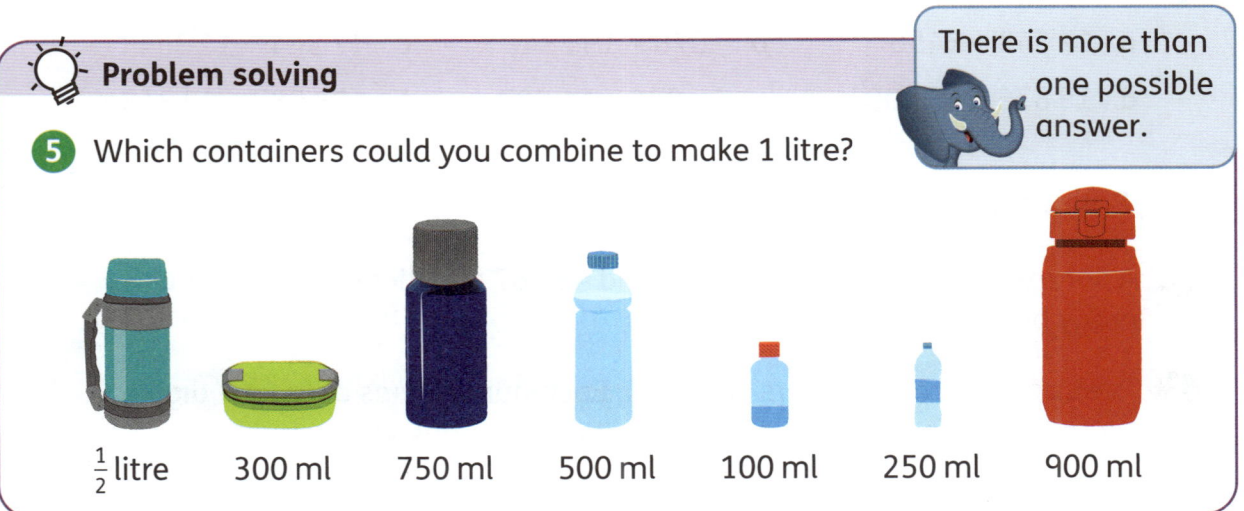

$\frac{1}{2}$ litre   300 ml   750 ml   500 ml   100 ml   250 ml   900 ml

Add larger numbers

You can use column addition to add large numbers. Work through each example. The place-value diagrams show the same calculations.

> Keep digits that are in the same places above each other.

42 876 + 7123

```
  42 876
+  7 123
  49 999
```

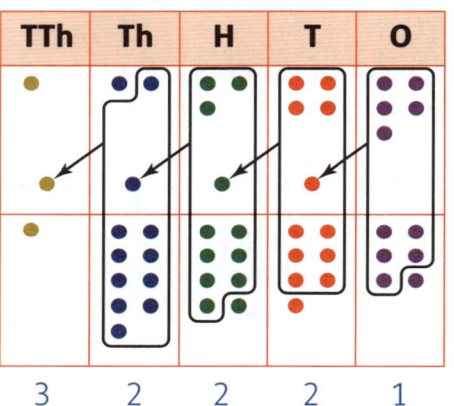

TTh	Th	H	T	O
4	9	9	9	9

12 345 + 19 876

```
  12 345
+ 19 876
  32 221
   1 1 1 1
```

TTh	Th	H	T	O
3	2	2	2	1

---

1 Add. Show your working. Draw place-value diagrams if you need to.

   a 12 543 + 6456   **b** 9876 + 19 234   **c** 12 344 + 11 354

   d 19 876 + 65 400   **e** 23 876 + 11 987   **f** 123 987 + 12 345

💡 Problem solving

2 A library has 42 945 fiction books and 45 987 non-fiction books. How many books does it have in total?

3 The sum of three numbers is 12 986. Each number has at least 3 digits. Write four different additions that will give this answer.

Subtract larger numbers

Look at these examples of column subtraction. The place-value diagrams show the same calculations.

12 459 − 1348

```
   12 459
 −  1 348
   11 111
```

TTh	Th	H	T	O
1	1	1	1	1

19 473 − 8534

```
   19 473
 −  8 534
   10 939
```

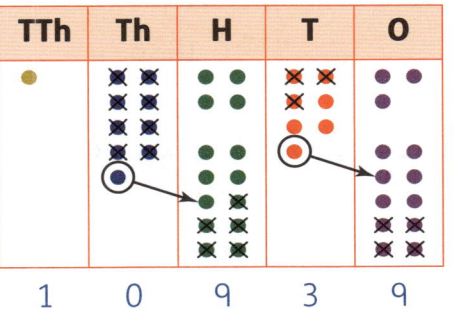

TTh	Th	H	T	O
1	0	9	3	9

Discuss how the models for subtraction on page 36 are different from the ones used for addition on this page. Why do they need to be different?

1 Subtract. Show your working.

  a 12 466 − 2314   b 19 324 − 17 103   c 23 450 − 4000

  d 17 312 − 9866   e 23 976 − 12 849   f 123 400 − 12 102

Problem solving

2 135 000 passengers used an airport in January to March this year.
19 465 fewer passengers used the airport in January to March last year.
How many passengers used the airport in January to March last year?

3 A football stadium has 78 250 seats. One match day, 57 896 seats were occupied. How many seats were not occupied?

Inverse operations

Look at these number sentences and the number lines that match them.

$45 + 32 = 77$ $77 - 32 = 45$ $32 + 45 = 77$ $77 - 45 = 32$

$439 + 897 = 1336$ $1336 - 897 = 439$ $897 + 439 = 1336$ $1336 - 439 = 897$

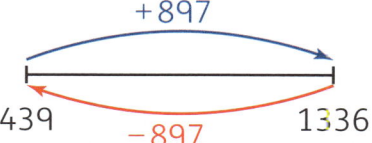
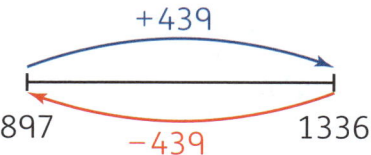

An **inverse operation** 'undoes' a calculation. How does this work with addition and subtraction?

How can the fact that addition and subtraction are inverse operations help you to check your answers when you add or subtract?

1 Use the fact that is given to find the missing number in the other fact.

    a $2345 + 7654 = 9999$ $9999 - 2345 = \boxed{}$

    b $12\,345 - 3456 = 8889$ $8889 + \boxed{} = 12\,345$

2 Use a written method to do each calculation. Then do the inverse operation to check your answer.

    a $12\,657 + 3456$ **b** $19\,349 + 13\,456$ **c** $12\,387 + 9876$

    d $14\,657 - 2146$ **e** $19\,345 - 12\,412$ **f** $24\,562 - 13\,483$

💡 Problem solving

3 Read the two statements about the bar model. Which statement is correct? Explain why.

19 417	
x	9214

    A x is 28 631 because you add 19 417 and 9214 to work out the inverse.

    B   The inverse operation is 19 417 subtract 9214, so x is 10 203.

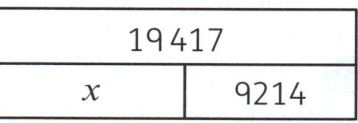

Workbook pages 22–23

Estimate and give approximate answers

An estimate is an approximate amount that is close to the real amount.

Estimating can help you decide whether the answer to a calculation is reasonable or not. You can use rounded numbers to estimate an approximate answer.

Look at this example:

> The numbers of visitors to an exhibition on three days were 5145, 5964 and 7499. Approximately how many visitors were there in total?
>
> Round the values to an appropriate place value.
> We will round these to the nearest 100.
>
> $5145 \rightarrow 5100$ $5964 \rightarrow 6000$ $7499 \rightarrow 7500$
>
> There were approximately 18 600 visitors.
>
> $$\begin{array}{r} 5\,100 \\ 6\,000 \\ +\ 7\,500 \\ \hline 18\,600 \end{array}$$

1 Estimate and then add. Use the method you find easiest.

a 345 + 795	**b** 1234 + 899	**c** 3199 + 999
d 1002 + 1765	**e** 988 + 1001	**f** 2999 + 4012

2 Estimate and then subtract. Use the method you find easiest.

a 602 − 141	**b** 425 − 299	**c** 500 − 299
d 6003 − 4597	**e** 4200 − 598	**f** 2800 − 1003

3 The Nile River is 6825 km long. The Congo River is 4375 km long. Approximately how much shorter is the Congo River than the Nile River?

4 Here are three calculations:

> 1276 + 4899   23 765 − 14 899   25 467 + 9587

a Round each number to the nearest 100. Work out an estimated answer to each calculation.

b Round each number to the nearest 1000. Work out an estimated answer to each calculation.

c What is the difference between your estimated answers?

d How does the place value that you round to affect the accuracy of your estimate?

Estimate when adding and subtracting

1 Work out an estimated answer to each addition.

    a 1529 + 1432   **b** 2584 + 864   **c** 2328 + 4618

    d 16 913 + 1101   **e** 25 055 + 12 599   **f** 12 361 + 17 085

    g 23 508 + 5008   **h** 44 090 + 25 110   **i** 38 999 + 31 999

2 Estimate each answer.

    a 2765 − 359   **b** 9997 − 480   **c** 8120 − 2099

    d 14 509 − 8999   **e** 23 412 − 12 459   **f** 38 999 − 12 345

Problem solving

3 A printer is printing 14 445 copies of a map of India and 9875 copies of a map of Asia. Each map fills one sheet of the printer's paper. Approximately how many sheets of paper will the printer need for this job?

> Always look at the numbers you are working with to decide how best to estimate.

4 An office sends 13 457 emails in one week and 11 248 in the next week. Approximately how many emails is this altogether?

5 Ria reads two stories. One story is 123 458 words long and the other is 99 543 words long. Approximately how many words does Ria read in total?

6 Arakan has 3465 photos on his phone. Asami has 4045 photos on her phone.

    a Approximately how many more photos does Asami have?

    b Arakan takes some more photos. After that he has 4817 photos on his phone. Approximately how many more photos did he take?

    c Approximately how many photos do Arakan and Asami have in total?

➡ *Workbook page 24*

Multi-step problems

Sometimes you need to do more than one calculation to solve a problem.

Think about this problem.

3817 people went to an exhibition. 1255 were adults under 50 years old. 1833 were 50 or over. The rest were children. How many children were there?

all visitors

3817		
1255	1833	?

under 50   50 or over   children

What calculations would you need to do to solve the problem?
How does the bar model help you to see what steps you need to take?

Marija says there were 719 children. Is she correct?
Do the calculations to check her answer.

Problem solving

Draw a bar model to show the information you are given.

1 A supermarket had 1750 bottles of water in stock on Monday. During the day, they sold 1089 bottles.
On Monday night, they got a delivery of 1140 bottles. How many bottles did they have in stock on Tuesday?

2 Kolya had £165. He bought a game for £29.99.
Then he spent £35.50 on clothes and £6.80 on lunch. How much money did he have left?

3 The areas of five large islands are given in the table.

Island	Madagascar	New Guinea	Borneo	Sumatra	Honshu
Area (km^2)	587 041	821 400	748 168	473 606	227 898

a With a partner, make up five multi-step problems using this data.

b Swap the problems with another pair. Solve the problems they have made up. Check each other's answers and discuss any mistakes.

➡ *Workbook page 25*

Decimals and percentages

Revisit decimal place value

💭 Think and share

Each grid represents one **whole**. What does one part of each grid represent?

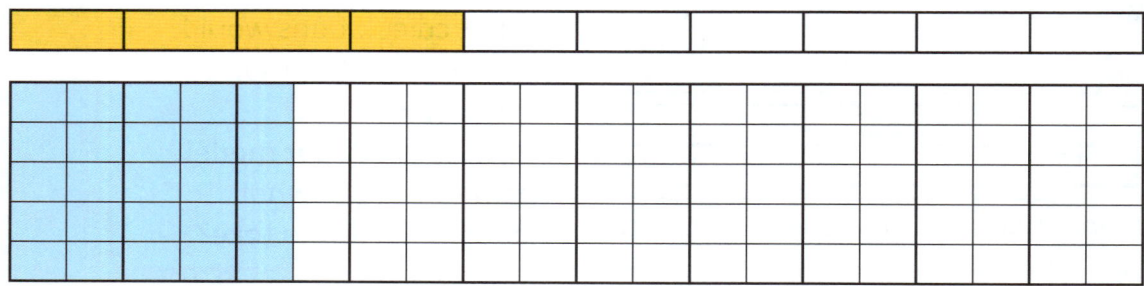

What is the shaded part of each grid as a fraction? What is it as a **decimal**?
What is the unshaded part of each grid as a decimal?

The place-value table shows that $\frac{25}{100}$ is made up of 2 tenths and 5 hundredths.

$\frac{1}{100}$ is made up of 0 tenths and 1 hundredth so we write a 0 in the tenths column as a place holder.

Ones	tenths	hundredths
0 •	2	5
0 •	0	1

1 For each diagram, write a decimal to represent the shaded part and another decimal to represent the unshaded part.

a	b	c	d	e	f

 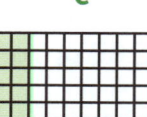 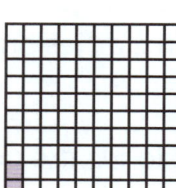 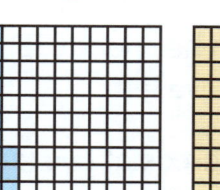

2 Look at the digit 9 in each number.

 47.9   9.13   0.39   3.96   90.6

a In which number does the digit 9 have the highest value?

b In which two numbers does the digit 9 have the same value?

➡ *Workbook page 26*

Thousandths

When you divide one whole into 10 equal parts, each part is $\frac{1}{10}$, or 0.1.
When you divide one whole into 100 equal parts, each part is $\frac{1}{100}$, or 0.01.

When you divide one whole into 1000 equal parts, each part is $\frac{1}{1000}$.
To write this as a decimal, you extend the place-value table to the right to include a place for **thousandths**.

Tens	Ones	•	tenths	hundredths	thousandths
2	4	•	2	5	7
1	6	•	0	4	1

$24.257 = 20 + 4 + \frac{2}{10} + \frac{5}{100} + \frac{7}{1000}$

$16.041 = 10 + 6 + \frac{0}{10} + \frac{4}{100} + \frac{1}{1000}$ The 0 is a place holder for tenths.

How is 16.041 different from 16.41?

1 Write each fraction as a decimal.

a $\frac{245}{100}$ b $\frac{999}{1000}$ c $\frac{99}{1000}$ d $\frac{105}{100}$

e $\frac{9}{1000}$ f $\frac{28}{1000}$ g $\frac{128}{1000}$ h $\frac{500}{1000}$

2 Look at the numbers.

a Which numbers have 4 in the thousandths place?

b Which number has the greatest number of hundredths?

c Which number could be written as: $1 + \frac{8}{10} + \frac{6}{100} + \frac{4}{1000}$?

8.164   6.148   1.864   1.486   8.461

Problem solving

Remember:
1000 m = 1 km

3 Marco travels 678 m to school. Kamala travels 0.68 km to school and Desi travels 0.02 km further than Marco.

a Who travels 0.678 km to school?

b Who lives closer to school, Desi or Kamala?

c How far does Desi travel to school?

➡ Workbook page 27

Represent decimals in different ways

Decimals can be represented in different ways, just like whole numbers.

Here are some representations of 35.267. Can you think of any other ways?

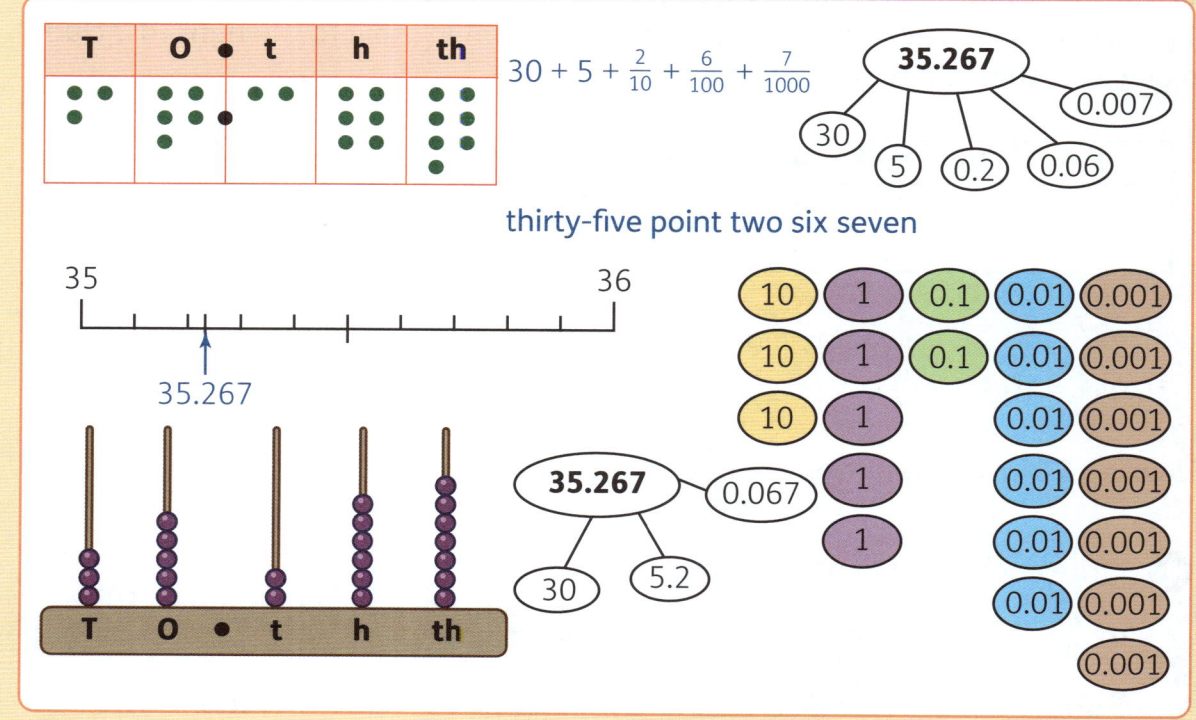

thirty-five point two six seven

1. Write each of these numbers as a decimal. Then represent the decimal in two different ways.

   a 9 ones and 4 tenths

   b 3 ones and 19 hundredths

   c $12\frac{3}{100}$

   d 45 hundredths

   e $\frac{9}{10}$

   f $\frac{9}{100}$

2. Write the decimals shown by the letters on each number line.

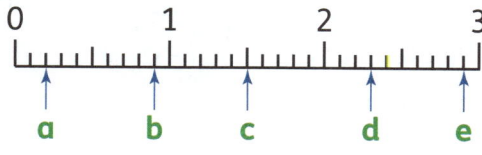

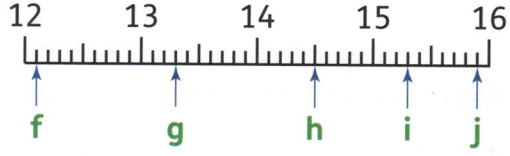

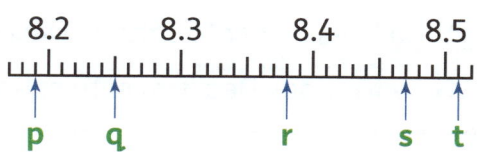

Compare and order decimals

You can compare decimals using place value, just as you compare whole numbers.

| Line up the decimal points. | ▶ | Compare digits, starting at the left. | ▶ | Stop at the first digit that is different. | ▶ | Compare these to decide which is greater. |

Fill in < or > to compare these numbers.

0.507 ☐ 0.529   0.14 ☐ 0.108

0.5**0**7   first place with
0.5**2**9   different digits
        0 < 2
    So, 0.507 < 0.529

line up
decimal points

0.14**0**   you can write 0
0.108   here as there
             are no thousandths

4 > 0

So, 0.14 > 0.108

Think:
$\frac{140}{100} > \frac{108}{100}$

---

1 Copy the statements. Fill in <, > or = to compare the decimals.

a 0.26 ☐ 0.52   b 0.13 ☐ 0.130   c 0.64 ☐ 0.46

d 3.4 ☐ 3.04   e 0.075 ☐ 0.75   f 0.385 ☐ 3.85

💡 **Problem solving**

2 A fairyfly is a small wasp 0.139 mm long. A featherwing beetle is 0.32 mm long.

a Which type of insect is longer?

b Is a fairyfly shorter or longer than $\frac{1}{10}$ of a millimetre?

c Another type of wasp is 10 times longer than the fairyfly. How long is it?

3 The mass of a beetle is a number with 3 decimal places between 2.608 grams and 2.615 grams. What could the mass be?

▶ *Workbook page 28*

Round decimals

You can round decimals to the nearest whole number or to a given decimal place using the rules for rounding.

> Round 1.25 to the nearest whole number.
> 1.25   the digit to the right of the whole number is 2, so the 2 doesn't change.
> 1.25 to the nearest whole number is 1.

> Round 1.25 to the nearest tenth.
> 1.25   the digit to the right of the tenths place is 5, so round up the 2.
> 1.25 rounded to the nearest tenth is 1.3.

1 Round each decimal to the nearest whole number.

    a 0.8   **b** 1.3   **c** 9.675   **d** 12.89   **e** 19.9

2 Round each decimal to the nearest tenth.

    a 164.21   **b** 234.73   **c** 765.04   **d** 543.47   **e** 599.91

3 Nathi rounded some masses to the nearest whole kilogram and got these results:

25 kg	26 kg	24 kg	20 kg	29 kg	30 kg

Copy the whole kilogram amounts. Then match the measurements in the box below to the whole kilogram amounts they round to.

28.5 kg	25.2 kg	24.5 kg	26.09 kg	29.8 kg	30.09 kg
25.6 kg	25.33 kg	24.98 kg	19.99 kg	20.35 kg	20.3 kg
25.9 kg	26.34 kg	29.45 kg	29.55 kg	25.8 kg	24.49 kg

 Problem solving

4 Some small mangoes have an average mass of 0.144 kilograms.

    a Is it correct to say the mangoes weigh about 0.1 kilograms? Why?

    b How many mangoes would you expect to get if you bought 1 kilogram? Why?

More decimals

1 Write the value of each underlined digit. Then write the next three numbers in each sequence.

   a 0.2<u>6</u>, 0.2<u>7</u>, <u>0</u>.28, ___, ___, ___

   b <u>1</u>.96, 1.<u>9</u>7, 1.9<u>8</u>, ___, ___, ___

   c 6.<u>2</u>5, 6.2<u>4</u>, <u>6</u>.23, ___, ___, ___

   d <u>8</u>.04, 8.0<u>3</u>, 8.<u>0</u>2, ___, ___, ___

2 Write the value of each underlined digit. Then write each set in order from smallest to greatest.

   a 2.<u>2</u>6, 1.<u>8</u>5, 2.<u>9</u>, 1.<u>3</u>5

   b 1.0<u>4</u>, 0.<u>4</u>0, 1.0<u>1</u>, <u>1</u>.4

   c <u>2</u>.79, 5.1<u>2</u>, 2.<u>1</u>7, <u>1</u>.38

   d 1.<u>2</u>5, 3.0<u>3</u>, 2.7<u>1</u>, <u>1</u>.48

3 Write these prices in order from greatest to smallest.

a

Apple juice	£1.39
Mineral water	£0.69
Yoghurt	£1.09
Bananas	£0.45
Passion fruit juice	£2.32
Satsuma juice	£1.38
Apricot halves	£0.52
Blackberries	£1.50
Half-fat milk	£0.30
Granary loaf	£0.65

b

Bathroom accessories:	
Gold-plated mirror	£22.99
Toothbrush holder	£9.50
Soap dish	£10.99
Shelf	£14.99
Towel rail	£14.95
Towel ring	£9.95
Toilet-roll holder	£9.99
Toilet-brush holder	£24.99

4 Rounding amounts can help you to work out a bill quickly.

   a Use the food bill from question 3a. Round all the prices. What is the approximate total cost?

   b Now use a calculator to work out the actual total. How close are the two answers?

Problem solving

5 The length of a piece of rope is 8.9 metres when rounded to the nearest tenth. Could the rope be shorter than 8.9 metres? Explain your answer.

Percentages

Per cent means 'in every hundred'.
The symbol % means 'per cent'.

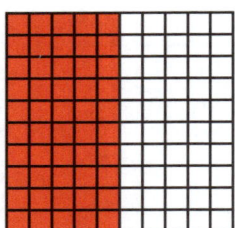 50% (or $\frac{1}{2}$) of this square is shaded.

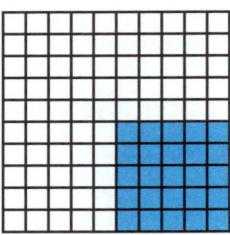 25% (or $\frac{1}{4}$) of this square is shaded.

1 What **percentage** of each square is shaded?

a 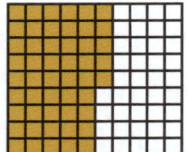 b 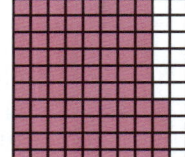 c  d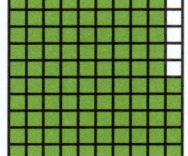

2 What percentage of each square is not shaded?

3 Richard answered 86 out of 100 questions correctly.
What percentage was this?

4 This is a waffle diagram. It shows the percentage of stars, hearts and squares on a sheet of stickers. Each symbol on the diagram represents 1% of the total number of stickers on the sheet.

The percentage of stars, hearts and squares on a sheet of stickers

a What percentage of the stickers is stars?

b What percentage of the stickers is neither stars nor squares?

c If there were 50 stickers on a sheet, how many of each type of sticker would there be? Explain how you got your answer.

5 Jasmine has 100 stickers. 63 stickers are butterflies.
What percentage of the stickers are not butterflies?

➡ *Workbook page 29*

Percentages, decimals and fractions

Percentages, decimals and fractions are different ways of showing the same value.

Look at the flowers.

3 out of the 6 flowers are red. We can say:

50% of the flowers are red.

0.5 of the flowers are red.

$\frac{1}{2}$ of the flowers are red.

The grid shows equivalent percentages, fractions and decimals.

50%					50%				
$\frac{1}{2}$					$\frac{1}{2}$				
0.5					0.5				
10%	10%	10%	10%	10%	10%	10%	10%	10%	10%
$\frac{1}{10}$	$\frac{1}{10}$	$\frac{1}{10}$	$\frac{1}{10}$	$\frac{1}{10}$	$\frac{1}{10}$	$\frac{1}{10}$	$\frac{1}{10}$	$\frac{1}{10}$	$\frac{1}{10}$
0.1	0.1	0.1	0.1	0.1	0.1	0.1	0.1	0.1	0.1

1. Copy and complete this table.

Percentage	1%	10%				
Decimal				0.75			0.98
Fraction			$\frac{6}{10}$		$\frac{83}{100}$	$\frac{9}{10}$	

2. Rewrite each set of numbers as fractions, in order from smallest to greatest.

a
0.6   50%
$\frac{27}{100}$ $\frac{3}{4}$

b
0.4 $\frac{1}{4}$
4% $\frac{1}{2}$

c
$\frac{3}{10}$ 0.9
0.03   33%

d
$\frac{1}{2}$ 0.2
0.02 $\frac{22}{100}$

e
15%   0.5
$\frac{5}{100}$ $\frac{3}{4}$

➡ *Workbook pages 30–31*

Find a percentage of an amount

Remember: a percentage is a fraction out of 100.

50% of an amount is $\frac{1}{2}$ of the amount.

25% of an amount is $\frac{1}{4}$ of the amount.

10% of an amount is $\frac{1}{10}$ of the amount.

1 A test had 12 questions.

    a Sam got 25% of the questions wrong.
How many questions did he get wrong?

    b Amira got 75% of the questions right.
How many questions did she get right?

    c Dan only answered 50% of the questions.
How many questions did he answer?

2 Nuresh has 30 marbles in a bag.

    a 50% of the marbles are blue. How many is that?

    b 20% of the marbles are red. How many is that?

    c 10% of the marbles are yellow. How many is that?

    d How many marbles are neither blue nor red nor yellow?
Tell your partner how you worked this out.

3 These are the ticket prices for a show at school.
Class 5 sold 200 tickets for the show.

TICKET PRICES	
Full price	$10.00
Student price	$7.00

    a 45% of the tickets were sold at student price.
How many of the 200 tickets is this?

    b The remaining 55% of the tickets were sold at full price.
How many tickets is this?

    c How much money in total should the class have collected
for the tickets they sold?

4 4175 pupils were asked which sporting events they would like to
watch at the Olympic Games. 84% of the pupils chose the 100 m
and 200 m athletics finals. Work out how many pupils chose
these events.

Equivalent percentages, fractions and decimals

You can write fractions and decimals as equivalent percentages using **equivalent fractions** with a **denominator** of 100.

$\frac{1}{4}$ of 100 pupils in a school wear glasses.

What percentage is this?

$\frac{1}{4}$ of 100 = 25 $\frac{1}{4} = \frac{25}{100}$, which is equivalent to 25% of the pupils.

Express the number of pupils who don't wear glasses as a decimal.

$\frac{1}{4}$ wear glasses, so $\frac{3}{4}$ do not.

$\frac{3}{4}$ of 100 is $\frac{75}{100}$, which is 75 hundredths, or 0.75 of the pupils.

1 Clothes labels tell you the percentage of each material used to make them.

Linen	50%
Cotton	☐

Cotton	☐
Polyester	25%
Nylon	12%

Cotton	☐
Lycra	3%

Linen	80%
Wool	15%
Cotton	☐

Cotton	☐
Linen	12%
Elastane	33%

Cotton	☐
Polyester	2%
Lycra	8%

  a Find the percentage of cotton in each item of clothing.

  b What fraction of each item is cotton?

2 The table shows the results of a school survey.

Question	Percentage who answered 'yes'
Do you go to bed before 10 p.m.?	50%
Do you watch more than 2 hours of TV per day?	75%
Do you have a pet?	60%
Do you have any siblings?	85%
Do you use a computer at home?	50%

Write a decimal to show what fraction of children:

  a go to bed before 10 p.m.

  b own a pet

3 What fraction of children:

  a watch less than 2 hours of TV per day

  b do not have a pet

  c use a computer at home

  d go to bed after 10 p.m.?

➡ *Workbook pages 32–33*

Time

Time zones

☁ **Think and share**

Suresh lives in Lahore in Pakistan. He wants to watch a cricket match that is being played in England. The match starts at 10 a.m. in England.

Use the map to explain why the match will start at 3 p.m. in Lahore.

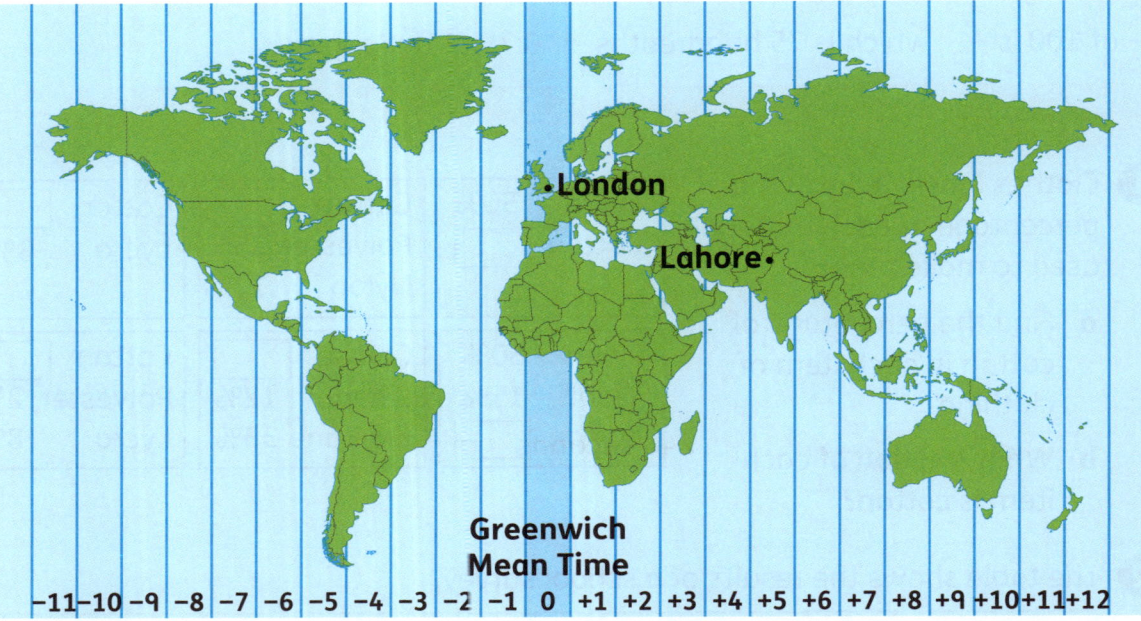

Greenwich Mean Time

| −11 | −10 | −9 | −8 | −7 | −6 | −5 | −4 | −3 | −2 | −1 | 0 | +1 | +2 | +3 | +4 | +5 | +6 | +7 | +8 | +9 | +10 | +11 | +12 |

← hours behind time in England
(earlier than England)

→ hours ahead of time in England
(later than England)

What time do you think it is in England right now? Why?

1 A football final starts at 5.15 p.m. in London on Monday. What time is this in these places?

    a Helsinki, Finland (2 hours ahead)   **b** Havana, Cuba (5 hours behind)

    c Mexico City, Mexico (6 hours behind)   **d** Tokyo, Japan (9 hours ahead)

    e Canberra, Australia (10 hours ahead)   **f** Nairobi, Kenya (3 hours ahead)

2 Tell your partner how you worked out each time.

Convert units of time

1 minute = 60 seconds	1 day = 24 hours	1 year = 52 weeks
1 hour = 60 minutes	1 week = 7 days	1 year = 12 months
$\frac{1}{2}$ hour = 30 minutes	1 year = 365 days	1 decade = 10 years
$\frac{1}{4}$ hour = 15 minutes	1 leap year = 366 days	1 century = 100 years

Multiply to convert from a larger unit of time to a smaller unit.

4 hours = ☐ minutes
$4 \times 60 = 240$ minutes

How many days in 12 weeks?
$12 \times 7 = 84$ days

Time units are not decimal because the units don't increase in multiples of 10.

Divide to convert from a smaller unit of time to a larger unit.

How many minutes is 180 seconds?
$180 \div 60 = 3$ minutes

How many years is 72 months?
$72 \div 12 = 6$ years

1 Convert between these units of time.

   a 3 hours → minutes   **b** 4 days → hours   **c** 8 hours → minutes

   d 2.5 days → hours   **e** 240 seconds → minutes   **f** 32 hours → days

2 How many days are there in:

   a 8 weeks   **b** 3 weeks and 2 days   **c** 10 weeks and 6 days?

3 How many weeks are there in:

   a 35 days   **b** 70 days   **c** 707 days?

4 Convert these months to years.

   a 36 months   **b** 240 months   **c** 126 months

Problem solving

5 How many weeks and days are there in 109 days?

6 Mish swam for 50 minutes every day for 10 days. How long did he swim for altogether?

Workbook page 34

Less than a second

The time it takes a world champion sprinter to run 100 m is measured in seconds. The measurements have to be very accurate, because there are fractions of a second between the different runners. We use decimals to show fractions of a second.

Usain Bolt set the world record for the 100 m sprint in 2009. He ran 100 m in 9.58 seconds.

This means that he took 9 seconds and $\frac{58}{100}$ of a second to run the race.

1. This table gives the results of a 400 m hurdles race.

Name	Time in seconds
Yasmani Copello (Turkey)	47.81
Alison dos Santos (Brazil)	46.72
Rai Benjamin (USA)	46.17
Abderrahman Samba (Qatar)	47.12
Karsten Warholm (Norway)	45.94
Kyron McMaster (BVI)	47.08

a What does .17 mean in 46.17 seconds?

b Write the race times in order from fastest to slowest.

c Who took $\frac{4}{100}$ of a second longer than Kyron McMaster to complete the race?

d The previous world record was 46.7 seconds. By how much did Karsten Warholm beat it?

These records were valid in 2021. Check they still stand today.

2. The male and female world records for different events are given in the table.

Event	Record holder and time	Record holder and time
100 m	Florence Griffith-Joyner 10.49 s	Usain Bolt 9.58 s
200 m	Florence Griffith-Joyner 21.34 s	Usain Bolt 19.19 s
400 m	Marita Koch 47.60 s	Wayde van Niekerk 43.03 s

a Work with a partner to write five problems using these times.

b Swap the problems with another pair. Work out the answers to each other's problems and then check each other's work.

➡ *Workbook page 35*

Calculate with times

What do you need to know in order to work out how long something takes?
How can you do a quick calculation?

You can count on or count back to solve problems involving time.

A cyclist started a race at 09:30 and finished 4 hours and 21 minutes later. At what time did they finish?

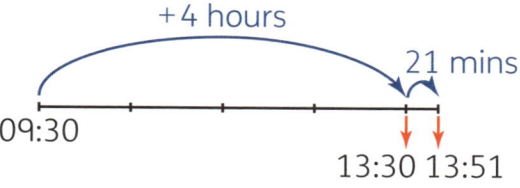

+4 hours
21 mins
09:30
13:30 13:51

They finished at 13:51.

Another cyclist started the race at 09:45 and finished at 14:57. How long did they cycle for?

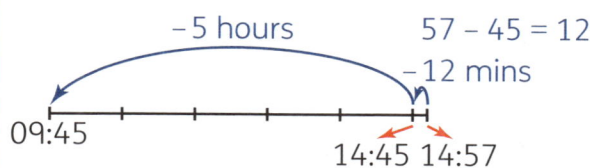

−5 hours   57 − 45 = 12
−12 mins
09:45
14:45 14:57

They cycled for 5 hours and 12 minutes.

1 The table shows the starting times, cycling times and finishing times of four cyclists.

Cyclist	Starting time	Cycling time	Finishing time
A	08:20	5 h 10 min	
B	10:05	6 h 2 min	
C	09:13		16:20
D	08:27		15:50

a Work out the missing times for each cyclist.

b The race officials started work at 08:15. They worked for $10\frac{1}{2}$ hours. At what time did they finish?

Problem solving

2 A cruise ship leaves Miami at 13:00 hours on 9 June. It returns 9.5 days later. Work out its return date and time.

3 A plane leaves Los Angeles at 11:45 to fly to London. The flight is 11.25 hours long. At what time will it arrive in London?

Los Angeles time is 8 hours earlier than London time.

➡ *Workbook page 36*

Mixed practice 1

1 Three numbers are represented here.

A Nine hundred and three thousand, seven hundred and ninety-eight.

B

HTh	TTh	Th	H	T	O
•• •	••• •	•• •		•• ••	••• ••• ••• •

C
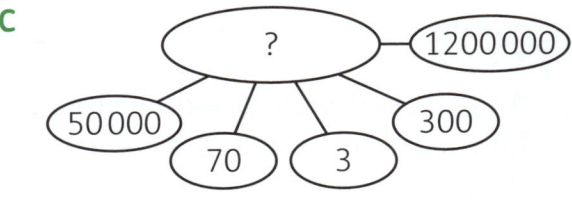

a Write each number in numerals.

b Underline the digit in the ten thousands place in each number.

c Write the numbers in descending order.

d Round number A to the nearest 1000.

e Round number B to the nearest 100 000.

f Write the number that is 15 000 less than each number.

g What do you need to add to number A to make 1 million?

h How can you get from number C to 1 million in a single operation?

2 Rewrite each calculation as a column addition or subtraction and fill in the missing digits.

a 26☐65 + 593☐ = 3☐402

b 2☐82☐ + ☐3 271 = 59 0☐6

c ☐6☐7 − 94☐ = 2☐91

d 1☐32☐ − 9☐36 = 50☐9

3 Write the number represented by each letter on the number line.

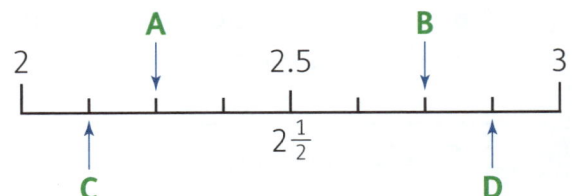

4 Draw a diagram to show the 3D shape you could make using this net.

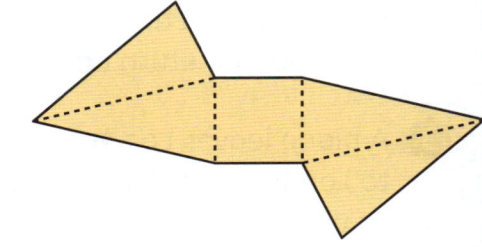

5 **a** What type of solid are these?

 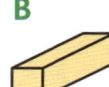 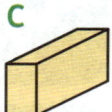

   b Draw a rough sketch of the net you would use to make each shape.

   c State one thing that is similar about the nets and one thing that is different.

6 Which of these numbers are between 4.5 and 5.4?

5.04   4.054   5.44   4.55   5.45   4.544   5.05   5.444   4.54

7 A mountain biker cycled along a trail for 1 hour 55 minutes. Then she stopped for $\frac{3}{4}$ of an hour to have a swim. She then took $1\frac{1}{4}$ hours to cycle back to the start of the trail. She started the trail at 07:40. At what time did she get back to the start?

8 Copy these angles accurately. Write the size in degrees and label each angle acute, right or obtuse.

a 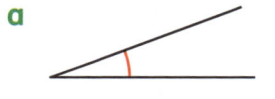   **b** 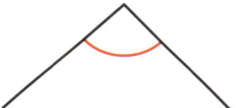   **c** 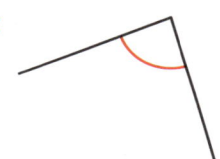   **d**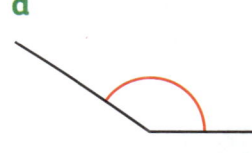

9 This waffle diagram shows how different pupils get to school.

   a What percentage of pupils walk to school?

   b If there are 260 pupils in the school, how many pupils use public transport to get to school?

   c Is it true that $\frac{1}{5}$ of the pupils get to school in private cars? Give a reason for your answer.

10 Make five pairs of equivalent times.

$2\frac{1}{2}$ hours	7200 seconds	$4\frac{1}{2}$ days	$2\frac{1}{2}$ days	20 minutes
120 minutes	150 minutes	108 hours	1200 seconds	3600 minutes

11 Calculate the sizes of angles a, b, c and d in this diagram.

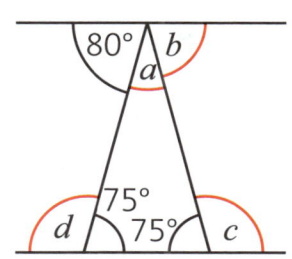

Multiplication and division 1

What do you already know?

💭 **Think and share**

What do we call arrangements in rows and columns like these?
How can they help you to work out multiplication and division facts?

Use the arrangements to make up five multiplication facts and five related division facts.

1 Work with a partner. Take turns to read a statement and say whether it is true or false.

   a 18 is a multiple of 6.   **b** 54 divided by 4 is 9.

   c 50 is a multiple of 10.   **d** 24 is a **factor** of 12.

   e 4 times 7 is 24.   **f** 1 times 1 is 1.

   g The product of 5 and 6 is 30.   **h** 30 divided by 3 is 10.

   i 45 divided by 5 is 8.   **j** 9 squared is 81.

   k 31 is a multiple of 2.   **l** 8 is a factor of 32.

2 Write the first six multiples of:

   a 8   **b** 9   **c** 12   **d** 7

3 Copy and complete these facts.

   a $6 \times \square = 54$ **b** $\square \times 3 = 36$ **c** $9 \times \square = 45$

   d $54 \div 6 = \square$ **e** $36 \div 3 = \square$ **f** $45 \div 9 = \square$

   g $5 \times \square = 60$ **h** $\square \times 8 = 56$ **i** $7 \times \square = 49$

➡ *Workbook pages 37–38*

Multiples

When you multiply two numbers, the product is a multiple of both numbers.

Write the first 12 multiples of 3 and the first 12 multiples of 5.

Multiples of 3: 3, 6, 9, 12, (15), 18, 21, 24, 27, (30), 33, 36
Multiples of 5: 5, 10, (15), 20, 25, (30), 35, 40, (45), 50, 55, (60)

All multiples of 3 can be divided by 3.
All multiples of 5 can be divided by 5.

What can you say about the circled multiples?

1 Look at this set of numbers.

24	28	48	56	72	45	70	49	54	36	32	60

Write the numbers from the set that are:

a multiples of 6
b multiples of 7
c multiples of 8
d multiples of 9
e multiples of both 6 and 8
f multiples of both 6 and 9
g multiples of both 8 and 9
h not multiples of 8 or 10

2 How can you tell quickly whether a number is a multiple of 10?

3 Copy and complete these number sequences.

a 12, 18, 24, __, __, __, __, __,

b 21, 28, 35, __, __, __, __, __,

c 81, 72, 63, __, __, __, __, __,

d 72, 64, 56, __, __, __, __, __,

4 List the first ten multiples of 6.

a Which of these are also multiples of 2?

b Which of these are also multiples of 3?

c Calculate 6 × 8 and then calculate 2 × 3 × 8. What do you notice? How can you explain this?

➡ *Workbook page 39*

Sequences of multiples

1 Write the next five numbers in each sequence. Copy and complete the sentences.

    a 125, 130, 135, 140
       These are all multiples of __.

    b 340, 350, 360, 370
       These are all multiples of 2, __ and __.

    c 425, 450, 475, 500
       These are all multiples of __ and __.

    d 550, 600, 650, 700
       These are all multiples of 2, __, __, __and __.

    e 200, 300, 400, 500
       These are all multiples of 2, 5, __, __, __, __ and __.

2 Use these rules to make number sequences.

    a List multiples of 25 between 801 and 901.

    b List multiples of 50 between 401 and 799.

    c List multiples of 100 between 1 and 999.

    d List numbers that are multiples of 10 and 25 between 101 and 249.

3 Write the term-to-term rule for each number sequence.

    a 20, 24, 28, 32, 36

    b 81, 27, 9, 3, 1

    c 250, 220, 190, 160, 130

    d 20, 200, 2000, 20 000

    e 128, 64, 32, 16, 8, 4

    f −18, −14, −10, −6, −2

4 Use your calculator to make up five new number sequences. Each sequence should have six numbers in it.

    a Write your number sequences on a piece of paper.

    b In your book, write the first term and the term-to-term rule for each sequence.

    c Swap your number sequences with a partner. Continue your partner's sequence for three more numbers.

    d Check each other's work. Tell your partner the term-to-term rules you used to find the next numbers in their sequences.

Use multiples to solve problems

Clare changes the oil in her motorbike every 3 weeks.
She changes her spark plugs every 5 weeks.

How often does she change the oil and spark plugs together?

First, list the multiples of 3 and 5.
Multiples of 3:   3, 6, 9, 12, 15, 18, 21, 24, 27, 30, ...
Multiples of 5:   5, 10, 15, 20, 25, 30, 35, ...

15 and 30 are multiples of both 3 and 5.
15 and 30 are **common multiples** of 3 and 5.
15 is the **lowest common multiple (LCM)** of 3 and 5.

Clare changes the oil and the spark plugs together every 15 weeks.

1 List the first ten multiples of 4 and 6.

   a   Which multiples are common multiples of 4 and 6?

   b   Which is the lowest common multiple?

2 Find the lowest common multiple of each pair of numbers:

   a   6 and 8   b   3 and 14   c   9 and 15

   d   4 and 5   e   5 and 7   f   3 and 9

3 In music, common multiples can help us to understand rhythms.
   Use common multiples to help you solve these problems.

   a   Sze Kui is learning to play
       the drums. Her teacher
       challenges her to play 3
       beats with her right hand
       in the same time as it takes
       her to play 2 beats with her
       left. Try this. Play the beats
       using your hands on the
       desk. On which beats do
       both hands play together?

   b   Now try playing 3 beats with one hand and 4 with the other hand.
       On which beats do both hands play together?

Square numbers

In Unit 3 you saw that some numbers can be organised into equal rows and columns to make a square.

9 and 25 are called square numbers.

A square number is formed when you multiply any number by itself.

When you multiply a number by itself, you say the number is **squared**.

The short way of writing 3 squared is 3^2 and the short way of writing 5 squared is 5^2.

$3 \times 3 = 9$

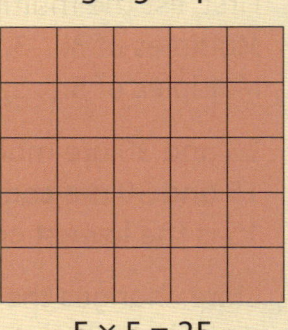

$5 \times 5 = 25$

1. Draw rough sketches of squares with blocks to show:

   a 4 squared   b 10 squared   c 6 squared   d 2^2

2. Work out the first 12 square numbers.

3. Copy these statements and fill in the missing numbers.

   a $9 = \square$ squared   b $1 = \square$ squared   c $4 = \square$ squared

   d $81 = \square$ squared   e $64 = \square$ squared   f $49 = \square$ squared

   g $144 = \square$ squared   h $100 = \square$ squared   i $16 = \square$ squared

   j $25 = \square$ squared   k $121 = \square$ squared   l $36 = \square$ squared

4. How is 4^2 different from 4×2? Discuss this with a partner.

Problem solving

5. A tiler has three packs of square tiles. One pack has 16 tiles, another has 24 and the third has 36.

   a Which of these packs can be laid out to form a square?

   b Can the tiler make a 5×5 square with the tiles he has? How?

   c What is the biggest square the tiler could make using tiles from all the packs?

Cube numbers

You can arrange **cube numbers** to make a cube.

$1 \times 1 \times 1 = 1$ $2 \times 2 \times 2 = 8$ $3 \times 3 \times 3 = 27$

The short way of writing $3 \times 3 \times 3$ is 3^3. Read 3^3 as 3 cubed. $3^3 = 27$.
27 is a cube number.

1 What cube numbers are shown here?

a
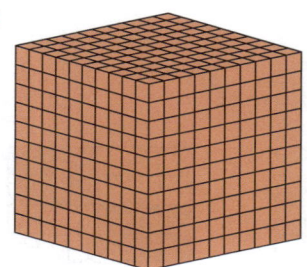
b

c
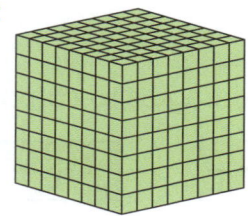
d
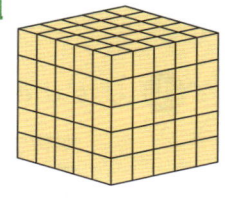

2 Is 16 a cube number? Give reasons for your answer.

3 Find the value of:

a 4^3 b 5^3 c 6^3 d $1^3 + 2^3$ e $1^3 + 2^3 + 3^3$

💡 Problem solving

4 Charlie does this drawing. He says that 360 is a cube number because it can be drawn as a cube. Is this correct? Give reasons.

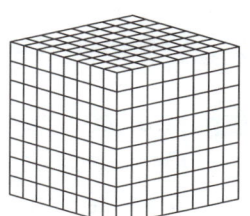

5 Read the clues and work out the numbers.

a I am the difference between 2^3 and 3^2.
What number am I?

b Half of me is the second cube number. What number am I?

c I am greater than 5^2 but less than 3^3. What number am I?

Factors

Some numbers cannot be organised into square shapes, but they can be organised into rectangles.

Look at these ways of organising the number 12. Each colour represents a factor.

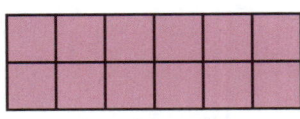  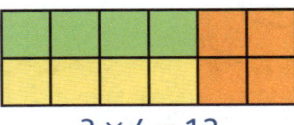

$1 \times 12 = 12$ $2 \times 6 = 12$ $3 \times 4 = 12$

12 has factor pairs 1 and 12, 2 and 6, and 3 and 4.
A factor is a number that divides exactly into another number.
The factors of 12 are 1, 2, 3, 4, 6 and 12.

1 Use these rectangles to complete the number sentences.

 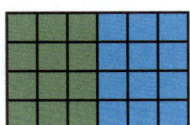 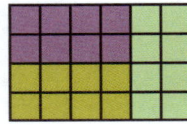 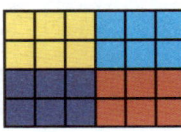

a $1 \times \square = 24$ $2 \times \square = 24$ $\square \times 8 = 24$ $\square \times 6 = 24$

b   The factors of 24 are: 1, 2, ____, ____, ____, ____, ____ and ____.

2 Write two pairs of factors for each of these numbers.

a 30	**b** 40	**c** 50	**d** 60
e 100	**f** 56	**g** 68	**h** 81

Problem solving

3 Four children throw counters onto this board.
One counter lands on a factor of 3 and of 12.
Another counter lands on a number that is not a factor of 12 or of 14.
The third counter lands on a factor of 18.
The last counter lands on a factor of 8.
Which number doesn't have a counter on it?

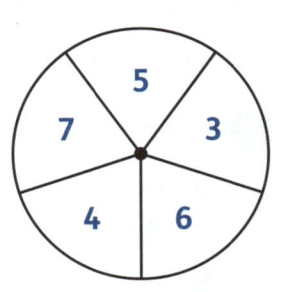

➡ *Workbook page 40*

Find factors of a number

Remember: the factors of a number can be multiplied to make that number. For example, the factors of 6 are 1, 2, 3 and 6.

$1 \times 6 = 6$ $2 \times 3 = 6$

A teacher has 24 students in her PE lesson.
She uses factors of 24 to group them for different exercises.
Here is an organised method for finding all the factors of 24.

Try 1: $1 \times 24 = 24$ 1 and 24 are factors of 24.
Try 2: $2 \times 12 = 24$ 2 and 12 are factors of 24.
Try 3: $3 \times 8 = 24$ 3 and 8 are factors of 24.
Try 4: $4 \times 6 = 24$ 4 and 6 are factors of 24.
Try 5: $5 \times ? = 24$ 5 is not a factor of 24.
Try 6:   We have already found that 6 is a factor of 24.

The factors of 24 are 1, 2, 3, 4, 6, 8, 12 and 24.

Can you think of any other ways to find all the factors of a number?

1 Find and list all the factors of each number.

   a 4   **b** 7   **c** 21   **d** 38   **e** 100

2 On squared paper, draw all the possible rectangles with area 36 squares.

Problem solving

3 There will be 40 guests at a dinner. All the tables must have the same number of people at them. What are the possible seating arrangements?

4 Use the diagram to answer the questions.

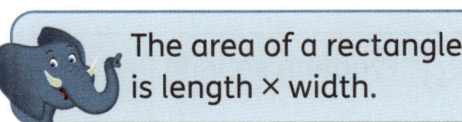

The area of a rectangle is length × width.

   a What is the area of the front of this envelope?

   b Find all the possible **dimensions** for envelopes with the same area.

   c Which of the answers to part b are practical for envelopes? Explain why.

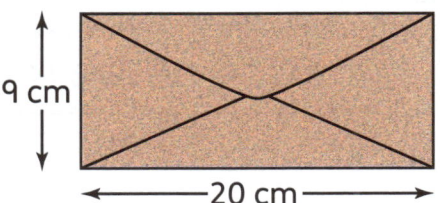

9 cm

20 cm

Common factors and multiples

You can divide both of these boxes into groups of 2, 3, or 6 crayons without any left over. What would happen if you used a different number?

30 pencils   30 pencils

Some numbers share factors. Here are two examples.
Factors of 24 are 1, 2, 3, 4, 6, 8, 12, 24.
Factors of 30 are 1, 2, 3, 5, 6, 10, 15, 30.
1, 2, 3 and 6 are **common factors** of 24 and 30.
6 is the **highest common factor (HCF)** of 24 and 30.

Some numbers share multiples.
Here are two examples.
Multiples of 4 are 4, 8, 12, 16, 20, 24, ...
Multiples of 6 are 6, 12, 18, 24, 30, ...

12 and 24 are common multiples of 4 and 6. The lowest common multiple (LCM) of 4 and 6 is 12.

1 List all the factors of each number. Find the highest common factor in each pair.

a   12 and 18   b   20 and 30   c   24 and 36

d   21 and 49   e   50 and 100   f   60 and 72

2 Find the lowest common multiple of each pair of numbers.

a   4 and 10   b   5 and 7   c   4 and 8

d   9 and 12   e   10 and 11   f   6 and 8

💡 Problem solving

3 Three children go to gymnastics club on 1st April. After that, Ani goes every third day, Misha goes every fourth day and Li goes every fifth day.
On what date will they next go on the same day?

➤ *Workbook page 41*

Prime numbers

Numbers with more than two factors are called **composite numbers**.

> 12 is a composite number. Its factors are 1, 2, 3, 4, 6 and 12.
> 25 is a composite number. Its factors are 1, 5 and 25. It is also a square number, because you can multiply one of its factors by itself to make 25.
> 1 is not composite. It only has one factor: 1.

Numbers with only two factors are called **prime numbers**.

> Each of these numbers has only two factors: 1 and the number itself. The number 1 is not a prime number because it has only one factor.
>
> Factors of 7:   1 and 7
> Factors of 11: 1 and 11
> Factors of 17: 1 and 17

1 One number in each set is not a prime number. Find this number and write its factors to show that it is not a prime number.

a   2 3 5 9

b   1 3 5 7

c   17 19 23 30

d   41 43 47 48

e   2 7 3
    5 11 21

f   71 73 75 79

g   97 95 89 83

h   13 17 18 19

i   89 91 97 99

2 Look at a calendar for this month. Write the dates that are prime numbers, for example 2nd, 3rd, 5th, …

Problem solving

A German mathematician called Christian Goldbach said that all even numbers greater than 2 can be written as the sum of two prime numbers. For example:

$4 = 2 + 2$ $6 = 3 + 3$ $8 = 3 + 5$

This is called Goldbach's conjecture.

3 Write these numbers as the sum of two prime numbers.

a 10   b 16   c 24   d 30   e 36

Multiply by 10, 100 or 1000

You can use facts you know and place value to help you multiply any number by 10, 100 or 1000.

TTh	Th	H	T	O
			9	1
		9	1	0
	9	1	0	0
9	1	0	0	0

- 91×10 – move the digits 1 place left; fill in 1 zero as a place holder.
- 91×100 – move the digits 2 places left; fill in 2 zeros as place holders.
- 91×1000 – move the digits 3 places left; fill in 3 zeros as place holders.

Each place on the place-value table is 10 times the value of the place to the right of it.

1 Multiply. Write the answers only.

  a 4×10 b 19×10 c 88×10

  d 132×10 e 400×10 f 987×10

  g 1098×10 h 8500×10 i 9999×10

2 Find the product.

  a 9×100 b 28×100 c 90×100

  d 425×100 e 800×100 f 762×100

  g 1208×100 h 1800×100 i 8050×100

3 Write the number that is 1000 times greater.

  a 19   b 87   c 412   d 3287

  e 300   f 1900   g 2405   h 999

4 One bag contains 12 apples. How many apples are there in 10 bags?

5 100 children each pay £23 for a school trip. What did they pay altogether?

6 How many months are there in 100 years?

7 A factory produces 2486 pieces of clothing per day. How many pieces does it produce:

  a in 10 days   b in 100 days   c in 1000 days?

➡ Workbook page 42

Divide by 10, 100 or 1000

$125 \times 100 = 12\,500$ $12\,500 \div 100 = 125$
$137 \times 1000 = 137\,000$ $137\,000 \div 1000 = 137$

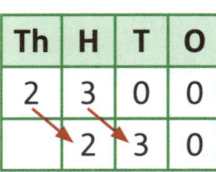
Division is the inverse of multiplication.

You use a place-value table to divide by 10 and 100.

Th	H	T	O
2	3	0	0
	2	3	0

$2300 \div 10 = 230$
Each digit moves 1 place to the right.

Th	H	T	O
	2	3	0
		2	3

$230 \div 10 = 23$
Each digit moves 1 place to the right.

Th	H	T	O
2	3	0	0
		2	3

$2300 \div 100 = 23$
Each digit moves 2 places to the right.

TTh	Th	H	T	O
2	3	0	0	0
			2	3

$23\,000 \div 1000 = 23$
Each digit moves 3 places to the right.

Each place on the place-value table is 10 times smaller than the value of the place to the left of it.

1 Do these divisions mentally. Write the answers only. Check them using a calculator.

a $670 \div 10$ **b** $800 \div 10$ **c** $420 \div 10$

d $9900 \div 10$ **e** $8760 \div 10$ **f** $2000 \div 10$

g $5000 \div 100$ **h** $3200 \div 100$ **i** $9900 \div 100$

j $23\,000 \div 1000$ **k** $198\,000 \div 1000$ **l** $1\,234\,000 \div 1000$

2 Find one tenth of:

a 340   **b** 590

c 800   **d** 3400

3 What is one hundredth of:

a 500   **b** 4400

c 9000   **d** 8900?

4 Roads on a map are $\frac{1}{10\,000}$ of their actual length.
A road is 900 m long in real life.
What length will it be on the map?

➡ *Workbook page 43*

Make multiplying simpler

You can use known facts and place value to multiply larger numbers mentally.

50×6

You know that $50 = 5 \times 10$.
$5 \times 10 \times 6$ is the same as
$5 \times 6 \times 10$
$= 30 \times 10$
$= 300$

400×8

You know that $4 \times 100 = 400$.
$4 \times 100 \times 8$ is the same as
$4 \times 8 \times 100$
$= 32 \times 100$
$= 3200$

1 Multiply. Try to work out the answers mentally.

a 3×20 b 6×30 c 5×80

d 7×40 e 6×80 f 2×70

g 9×60 h 5×50 i 6×40

j 8×80 k 9×90 l 4×90

2 Multiply. Try to work out the answers mentally.

a 200×4 b 200×8 c 9×200

d 300×5 e 6×300 f 9×400

g 7×500 h 600×8 i 900×7

j 800×9 k 900×2 l 700×8

3 Write how many animals there are altogether in:

a 6 flocks of 90 seagulls

b 3 schools of 400 fish

c 8 nests of 800 ants

d 6 herds of 80 elephants

e 9 groups of 400 antelopes

f 7 swarms of 800 locusts

g 4 groups of 50 caterpillars

h 400 nests with 4 birds in each

Write a number as a product of its factors

Writing a number as the product of its factors can make multiplication easier.

6×13 $\qquad 6 = 2 \times 3$ $= 2 \times 3 \times 13$ $3 \times 13 = 3 \times 10 + 3 \times 3 = 30 + 9 = 39$ $2 \times 39 = 2 \times 30 + 2 \times 9 = 60 + 18 = 78$	16×5 $8 \times 2 \times 5 \qquad\qquad$ or $4 \times 4 \times 5$ $= 8 \times 10 \qquad\qquad\quad = 4 \times 20$ $= 80 \qquad\qquad\qquad\quad = 80$

1 Complete the number sentences to show how you can use factors to do each multiplication.

a 14×7

$= 2 \times \square \times 7$

$= 2 \times \square$

$= \square$

b 12×8

$= 2 \times \square \times 8$

$= 2 \times \square$

$= \square$

c 6×17

$= 3 \times \square \times 17$

$= 3 \times \square$

$= \square$

d 9×23

$= 3 \times \square \times 23$

$= 3 \times \square$

$= \square$

e 6×15

$= 2 \times \square \times 15$

$= 2 \times \square$

$= \square$

f 4×19

$= 2 \times \square \times 19$

$= 2 \times \square$

$= \square$

2 Multiply.

a 6×23

b 8×19

c 12×13

d 9×15

e 6×31

f 12×22

Choose the easier number to write as a product of its factors.

 **Problem solving**

3 Eggs are packed in trays of 30. How many eggs are there in 24 trays?

Order of operations

Ms Chetty asked her class to find the answer to this problem:
$24 + 6 \div 2 - 1 \times 4$

This is how four pupils worked.

Chandra	Mohinder	Aneesa	Jay
$24 + 6 \div 2 - 1 \times 4$	$24 + 6 \div 2 - 1 \times 4$	$24 + 6 \div 2 - 1 \times 4$	$24 + 6 \div 2 - 1 \times 4$
$= 30 \div 2 - 1 \times 4$	$= 30 \div 2 - 4$	$= 24 + 3 - 1 \times 4$	$= 24 + 3 - 1 \times 4$
$= 15 - 1 \times 4$	$= 15 - 4$	$= 24 + 3 - 4$	$= 24 + 2 \times 4$
$= 14 \times 4$	$= 11$ ✗	$= 27 - 4$	$= 26 \times 4$ ✗
$= 56$ ✗		$= 23$ ✓	$= 104$

Aneesa got the right answer because she did the different parts of the calculation in the correct order.

In mathematics there are order of operation rules for working things out. The three main rules are:
- Do the parts in brackets first.
- Do the multiplication and division next. Work from left to right to do this.
- Do the addition and subtraction next. Work from left to right to do this.

These rules make sure that everyone solves problems with more than one operation in the same order and gets the correct answers.

Work out what mistakes Chandra, Mohinder and Jay made.

1. Are these statements correct? If not, say why not.

    a For $3 + 5 \times 2$, I would work out $3 + 5$ first.

    b To get $10 - 3 \times 5$, I would subtract 3 from 10 first.

    c To work out $6 + 12 \times 3$, I would first multiply 12×3.

    d For $8 + 9 \div 3$, I would work out $9 \div 3$ first. Then I would add.

    e To find $4 + 5 + 6 + 10$, I could add in any order.

    f For $(9 + 6) \div 5 + 2$, I would add $9 + 6$ first.

➡ *Workbook page 44*

Work in order

1 Follow the order of operation rules. Show all the steps in your working.

a $18 + 8 \times 2$ **b** $(18 + 8) \times 2$ **c** $26 - 4 \times 2$

d $(26 - 4) \times 2$ **e** $24 \div 4 + 3$ **f** $28 \div (4 + 3)$

g $49 - (21 \div 3)$ **h** $(49 - 21) \div 2$ **i** $23 - (6 \times 3)$

2 Calculate. Show your working.

a $14 + 8 - 11$ **b** $45 \div 9 \times 3$ **c** $13 - 20 \div 5$

d $20 \times 30 + 23$ **e** $25 \times (3 - 2)$ **f** $40 \times 50 \div 100$

g $104 + 4 \div 2$ **h** $88 \div 8 + 12$ **i** $45 - 23 + 29$

j $80 - 36 \div 3 \times 5 + 24 \div 8$

k $50 + 3 \times 2 + 18$

l $42 - 42 \div 6 \times 5 + 24 \div 3$

m $(18 - 2) \times (8 + 12)$

3 Calculate.

a $12 \div (4 + 2) \times 3$ **b** $20 - 3 - (2 + 7)$ **c** $8 \times 4 \div (8 \div 2)$

d $24 - (6 - 2) + 8$ **e** $24 + 2 - 3 + 4$ **f** $10 + 2 \times 8 + 4$

g $(12 - 4) \times (8 - 5)$ **h** $40 \div 4 - 24 \div 6$ **i** $18 + 8 - 7 + 3$

Problem solving

4 Kassim has the correct answers, but he has crossed out some of the operations. Work out what the missing operations are in each calculation. Rewrite the calculations correctly in your book.

a $2 + 21 \rule{1cm}{0.3mm} 3 = 9$ b $5 \times 3 \rule{1cm}{0.3mm} 8 = 7$

c $15 - 6 \rule{1cm}{0.3mm} 2 = 12$ d $14 - 8 \rule{1cm}{0.3mm} 6 = 0$

e $9 \rule{1cm}{0.3mm} 6 + 10 = 13$ f $12 \times 4 \rule{1cm}{0.3mm} 6 = 8$

g $18 \rule{1cm}{0.3mm} 8 - 2 = 8$ h $36 \rule{1cm}{0.3mm} 6 + 2 = 8$

i $15 \div (3 \rule{1cm}{0.3mm} 5) + (3 \times 5) = 16$ j $6 \rule{1cm}{0.3mm} 12 \div (3 \rule{1cm}{0.3mm} 1) = 9$

Measures and money

Length

> ### 💭 Think and share
>
> How tall do you think a giraffe is?
>
> Dimensions of a giraffe, on average:
> legs: 1780 cm long
> neck: 1.8 m
> head: 515 mm + 12.7 cm horns
> body height from top of leg to shoulder: 170 cm
>
> Remember:
> 1 metre (m) = 100 centimetres (cm)
> 1 centimetre = 10 millimetres (mm)
> 1 kilometre (km) = 1000 metres
>
> - Work in groups.
> - Read the information and work out the height of an average giraffe.
> - Tell the class how you did this.

1 Measure the lengths marked on this drawing in millimetres.
Record the measurements in millimetres and write the equivalent measurements in centimetres.

2 List five things that are usually measured in:

  a kilometres   b metres

  c centimetres   d millimetres

3 Write these lengths in centimetres.

  a 1 cm 1 mm   b 4 cm 5 mm   c 1 m 28 cm

  d 3 m 28 cm 4 mm   e 26 m   f 26.7 m

➡ *Workbook page 45*

Mass

We measure mass in milligrams, grams, kilograms and tonnes.
1 gram (g) = 1000 milligrams (mg)
1 kilogram (kg) = 1000 grams
1 tonne (t) = 1000 kilograms

1 Match each item to a suitable mass from the box.

115 g	11.6 t	36 g	25 kg	3442 kg	500 mg

a A glue stick

b A small paperclip

c A truck with a load of bricks

d A sack of potatoes

e A van full of potatoes

f A toy phone

2 Write these amounts in grams (g).

a 5 kg

b 3.5 kg

c 1 kg 143 g

d 7.2 kg

e 45.5 kg

f 11.7 kg

3 Order these masses from heaviest to lightest.

5 tonnes	250 g	5298 g	4.5 kg	1 kg
980 g	12 876 g	5200 kg	92 500 g	20 g

Capacity

Capacity is a measure of how much liquid a container can hold. Capacity can be measured in litres (ℓ) and millilitres (ml).

A millilitre is a small amount. A teaspoon holds about 5 millilitres of liquid.

1 litre = 1000 ml   0.5 litres = 500 ml

1 The capacity of each container is given but the units have been left off. Write the measurements with the correct units.

> The containers are not drawn to the same scale.

a 250 ☐

b 15 ☐

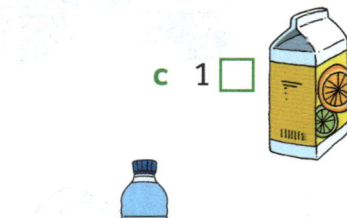

c 1 ☐

d 750 ☐

e 1.5 ☐

f 500 ☐

g 25 ☐

2 Write these litre amounts in millilitres.

a 1 litre   **b** 5 litres   **c** 10 litres

d 12.4 litres   **e** 4.5 litres   **f** 15.8 litres

3 Write these capacities in order from smallest to greatest.

3 ℓ	4.5 ℓ	2500 ml	8000 ml	3 ℓ and 250 ml
976 ml	1200 ml	1.6 ℓ	4 ml	90 ml	0.9 ℓ

Measuring scales

A **measuring scale** is the set of marks on a measuring instrument.
Usually only the main units are marked and numbered. You have to work
out what the in-between marks represent.

Here is a section of a ruler.

The scale is marked in centimetres.
Each centimetre is numbered.

- What are the centimetres divided into on this ruler?
- What does each small mark between the centimetres represent?
- How long is the red line? Give the measurement in two ways.

1 Here is a measuring jug.

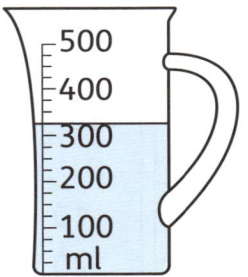

   a What units are shown on the scale?

   b What do the numbers on the scale tell you?

   c What does each small mark between the
      hundreds represent?

   d How much liquid is in the jug?

2 Work with a partner. Say what units are shown on each scale and
  what each mark represents.

a

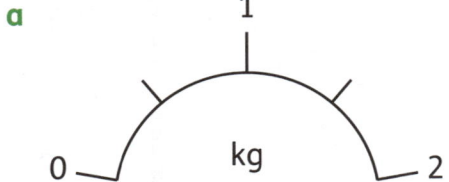

b

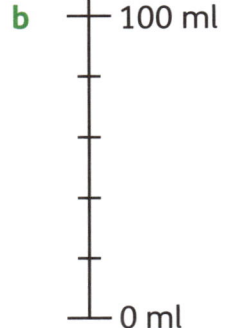

💡 **Problem solving**

3 Design five different scales to measure a capacity of 1 litre. The scales
  should be divided into 2, 4, 5, 8 and 10 equal parts.

Read measuring scales

1 Write the measurements shown by the letters on each scale.
Remember to write the units as well.

a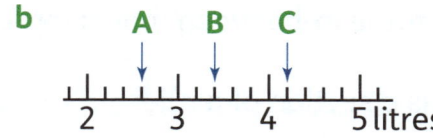

b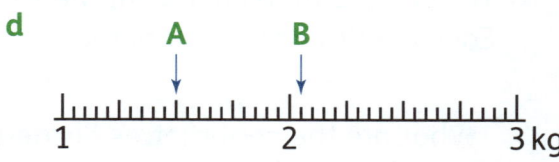

c

A   B

| | | | | |
2   3   4   5   6 cm

d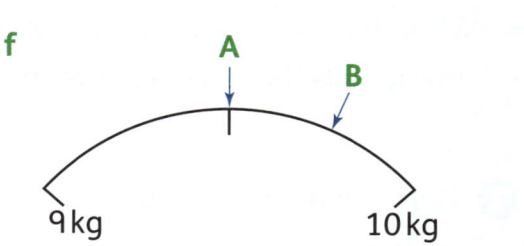

e

°C
50
A → 40
30
B → 20
10
C → 0
−10
−20
−30

f

A   B

9 kg   10 kg

g

metres
1.5
1
A → 0.5

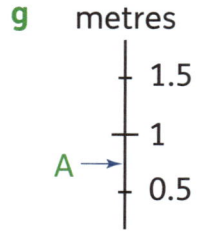

h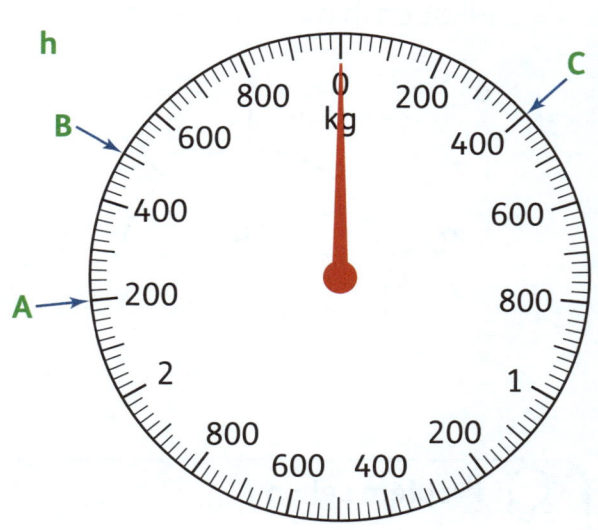

➡ *Workbook pages 46–47*

Metric and imperial equivalents

Even though most countries use the **metric system** for measurements, people still talk about length, mass and capacity in older **imperial** units.

Have you heard or seen any of these units being used?

You can use approximate measures to convert between metric and imperial units.

1 metre is approximately 3 feet.
1 kilogram is approximately 2 pounds.
1 litre is approximately 2 pints.
1 inch is approximately 2.5 cm.
1 ounce is approximately 30 grams.

Airport 17 miles

Spring Creek depth: 81 feet

49 p per $\frac{1}{2}$ pound

1 pint

1 Which of these are not metric units? Say what they are used to measure.

pound	gallon	kilogram	gram	pint	foot
stone	ounce	milligram	inch	litre	mile

2 Which is greater?

a 3 feet or 3 metres   **b** 2 pints or 2 litres   **c** 5 grams or 5 ounces

d 4 pounds or 4 kilograms   **e** 3 inches or 3 centimetres

3 Order these heights from shortest to tallest.

5 feet 4 inches   180 centimetres   6 feet 1 inch   1.75 metres

4 Approximately how many grams are in:

a 2 ounces   **b** $3\frac{1}{2}$ ounces   **c** 10 ounces?

5 Look at this graph carefully.

a What does the graph show?

b Use the graph to convert 40 km to miles.

c How many kilometres is 50 miles?

d Which is further, 60 km or 45 miles?

e Is it true that 8 km is about 5 miles?

Conversion graph: km to miles

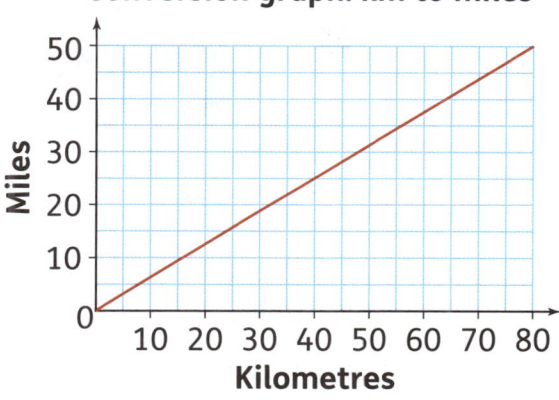

➡ *Workbook page 48*

Solve measurement problems

1 This week three gardeners used $43\frac{1}{2}$ litres, 41.3 litres and 41 000 ml to water plants. How much water is that altogether?

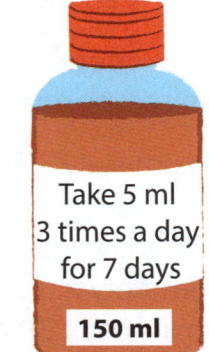

When you solve problems with measurements, convert units so they are all the same.

2 A patient has this bottle of medicine. How much medicine will be left in the bottle after 7 days?

Take 5 ml 3 times a day for 7 days

150 ml

3 Maryam bought 12.5 litres of juice for a party. She used $9\frac{1}{4}$ litres. How much is left?

4 A tailor needs 60 metres of blue fabric. He has four lengths in stock: 500 cm, 900 cm, 13.2 m and 9.7 m. How much more does he need to order?

5 Sal used ribbon to decorate a hall. She used 9 times more red ribbon than gold ribbon. She used 225 metres of red ribbon. How much gold ribbon did she use?

6 Here are the masses of different animals.

195 kg

1450 kg

803 kg

5099 kg

a What is the combined mass of an elephant and a giraffe?

b What is the difference in mass between the heaviest and lightest animal?

c A game ranger weighs 82 kilograms. How much lighter is she than a lion?

Work with money

1 Zara has these coins.

a Write five different amounts she could pay using two of these coins.

b Make a list of possible amounts that she could pay using three of these coins.

c She pays for an item costing £1.29 using one coin. Which coin does she use? What change does she get?

d Which coins could Zara use to pay each of these amounts exactly?

£1.60   £2.80   £0.70

2 Here are the prices at a café.

Work out how much change you get from a £10 note when you buy:

a a roti and a drink

b noodles, a sandwich, a roti and 3 drinks

c a sandwich, chips and a drink

d 6 orders of chips

MENU	
Roti	£2.40
Sandwich	£2.35
Chips	95p
Noodles	£1.80
Drink	99p

Problem solving

3 What is the total cost of 1.5 kg of mixed dried fruit at £5.80 per kilogram, $2\frac{1}{4}$ kg of raisins at £1.55 per kilogram and 500 g of dried apricots at £6.50 per kilogram?

4 Which pack of dates is the best value for money? Explain why.

0.25 kg
£2.25

450 g
£3.30

1 kg
£7.76

➤ Workbook page 49

Perimeter

Think and share

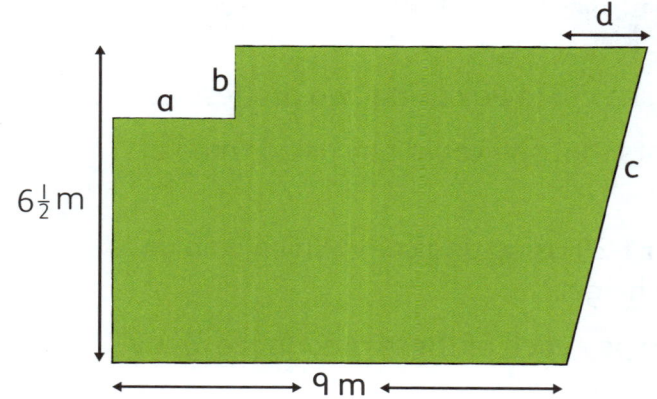

Mia wants to measure the **perimeter** of her lawn. She says:

> You don't need to know the length of sections a and b, but you do need to know the length of sections c and d.

Why is Mia correct?

Section d is $1\frac{1}{2}$ m long and section c is 6.7 m long.

How would you work out the perimeter of the lawn using these measurements?

Think of another way to work out the perimeter of the lawn. Share your ideas with your group.

1 Calculate the perimeter each shape in centimetres.

 The shapes are not drawn to scale, so you cannot measure them.

a
12.6 cm
4 cm

b
19 mm   19 mm
19 mm

c
41 cm
37 cm   37 cm
41 cm

2 The perimeter of each shape is given. Work out the missing lengths.

a
23 mm   23 mm
18 mm   18 mm
Perimeter = 97 mm

b
28 mm   28 mm
Perimeter = 112 mm

c
10.1 cm
10.1 cm
Perimeter = 24.6 cm

➤ Workbook page 50

Composite shapes

Composite shapes are made by combining shapes or by removing parts of shapes.

Describe each of these composite shapes in two different ways.

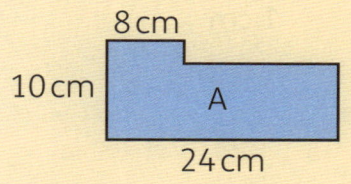
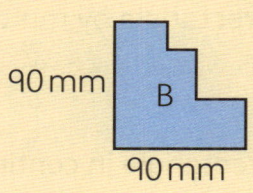
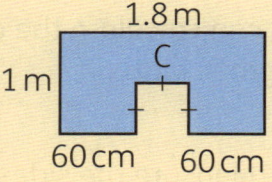

How can you use the properties of squares and rectangles to work out the lengths of sides that are not labelled?

What are the lengths of the unlabelled sides in each shape?

1 Calculate the perimeter of each shape.

> Convert measurements to the same units before you add them.

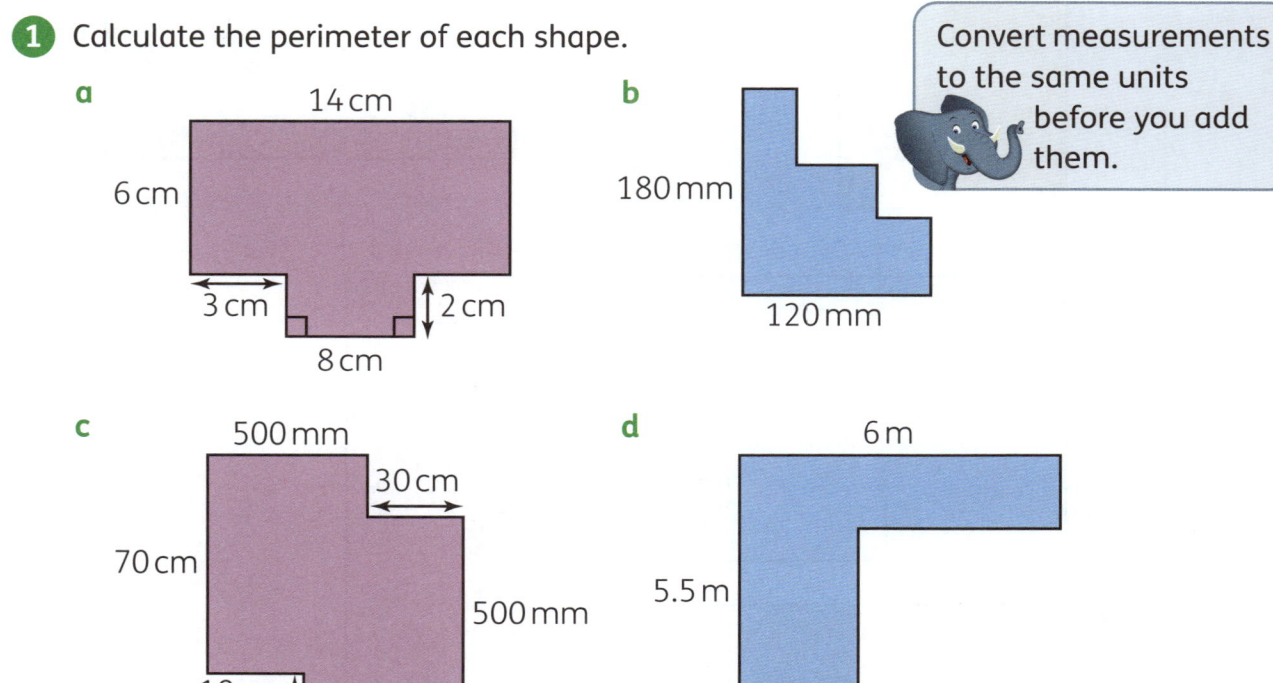

Problem solving

2 A regular pentagon has perimeter 29 metres. What are the lengths of the sides?

3 What is the cost of fencing a rectangular field with side lengths of 32 m and 19.5 m, with fencing that costs £42.00 per metre?

Area

Area is the amount of space covered by a shape.

The area of this square is one square centimetre (1 cm^2).

1 cm
1 cm

You can calculate the area of a rectangle by multiplying length × width.

1 Measure the sides of each rectangle in centimetres. Then calculate its area.

a
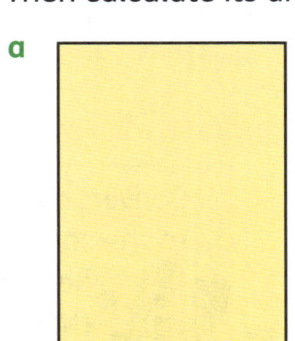

b

c
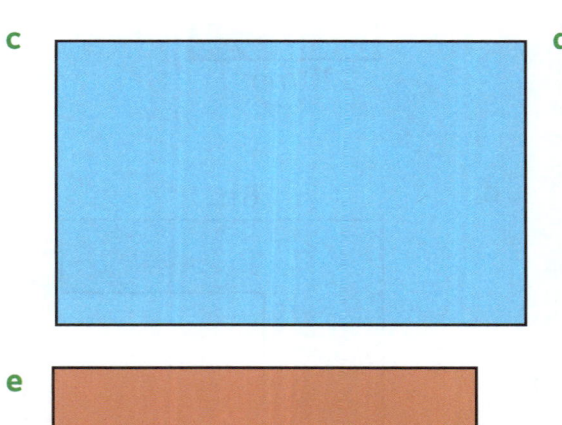

d

e

 Problem solving

2 A rectangle has area 24 cm^2. What could its side lengths be?

3 A square has a perimeter 36 cm. What is its area?

Use a formula to calculate area

The **formula** for area of a rectangle is
Area = length × width
$A = l \times w$

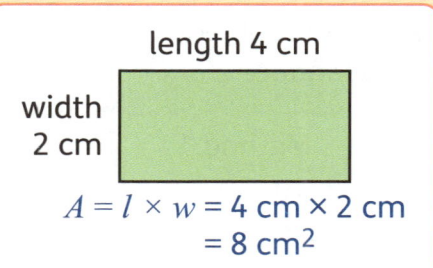

$A = l \times w = 4 \text{ cm} \times 2 \text{ cm}$
$= 8 \text{ cm}^2$

The sides of a square are all the same length.
The formula for area of a square is
Area = side × side
$A = s \times s$, or $A = s^2$

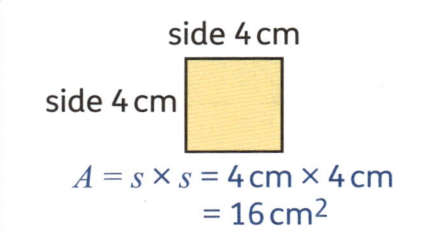

$A = s \times s = 4 \text{ cm} \times 4 \text{ cm}$
$= 16 \text{ cm}^2$

1 Use the formula to calculate the area of each of these rectangles.

The rectangles are not drawn to scale.

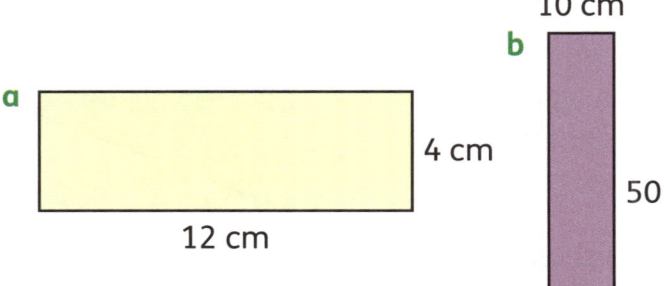

a   12 cm, 4 cm

b   10 cm, 50 cm

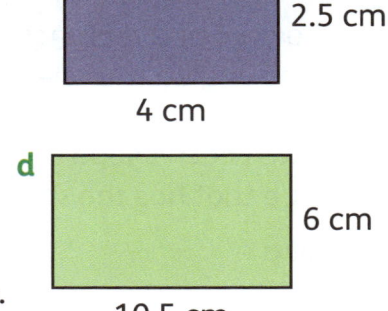

c   4 cm, 2.5 cm

d   10.5 cm, 6 cm

2 Here is a table of dimensions for some rectangles.
Copy it and fill in the missing information.

Length	Width	Area
5 cm	3 cm	
12 cm	2 cm	
9 cm		36 cm²
11 cm	8 cm	
7 cm	7 cm	
15 cm		90 cm²
	6 cm	72 cm²
9.5 cm	4 cm	

3 Find some rectangles in the classroom. Measure them and find their area.

More area calculations

Work through each method to see how three pupils found the area of the same shape.

Method 1

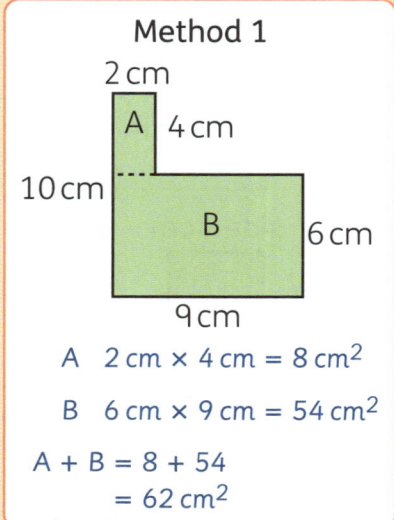

A $2\,cm \times 4\,cm = 8\,cm^2$

B $6\,cm \times 9\,cm = 54\,cm^2$

$A + B = 8 + 54$
$\qquad = 62\,cm^2$

Method 2

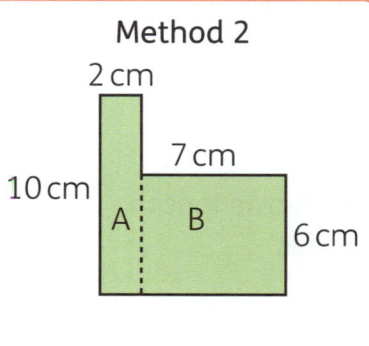

A $2\,cm \times 10\,cm = 20\,cm^2$

B $7\,cm \times 6\,cm = 42\,cm^2$

$A + B = 20 + 42$
$\qquad = 62\,cm^2$

Method 3

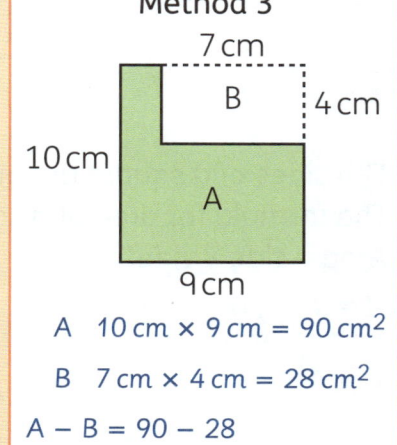

A $10\,cm \times 9\,cm = 90\,cm^2$

B $7\,cm \times 4\,cm = 28\,cm^2$

$A - B = 90 - 28$
$\qquad = 62\,cm^2$

Which of these methods do you like best? Why?

1 Use this diagram to explain how you can use the area of a rectangle to find the area of a right-angled triangle.

2 Match each shape in Set a with the shape from Set b that has the same area.

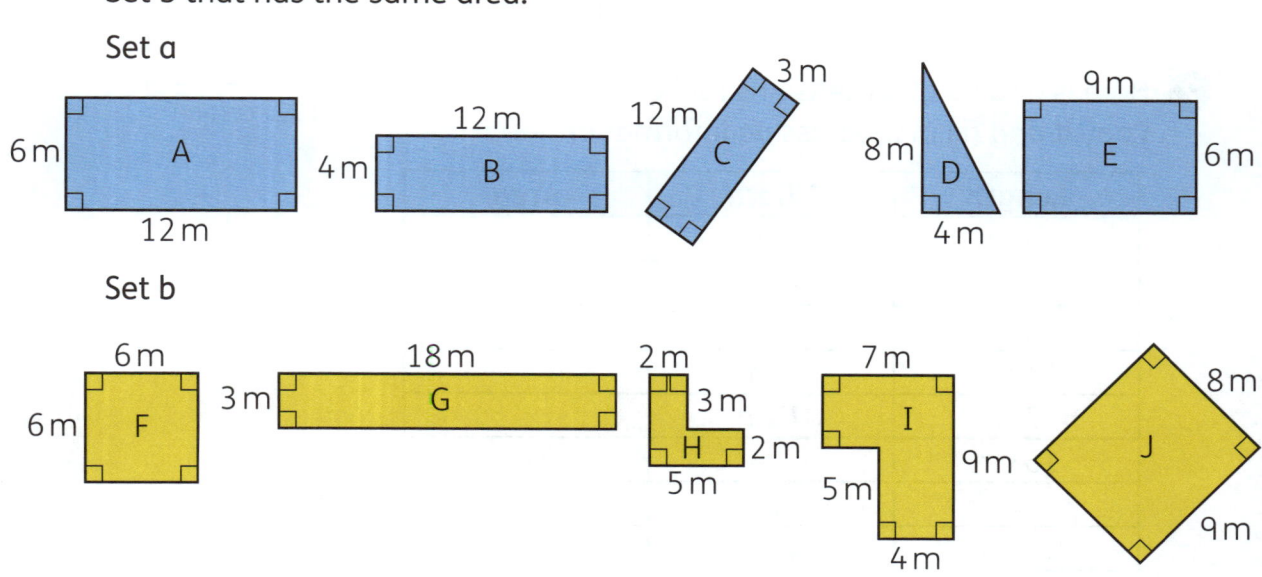

Set a

Set b

▶ Workbook page 51

Solve perimeter and area problems

1 Find the area and perimeter of each of these shapes.

a

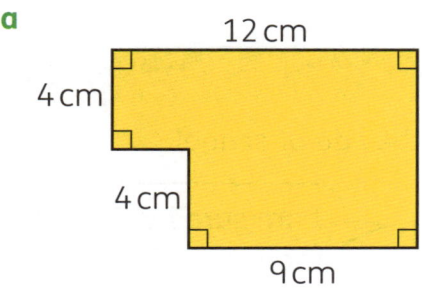

b

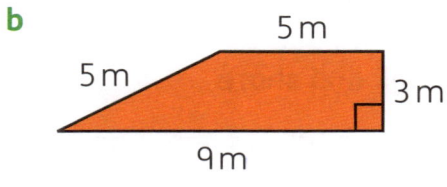

c

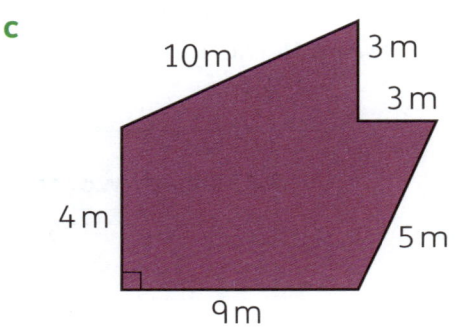

> When you solve problems involving shapes and measurements, always do a rough sketch and label it.

2 How many different rectangles can you make with a perimeter of 60 cm? The side lengths should all be whole numbers.

3 Which has the greater area: a square with sides of 120 mm or a rectangle with sides of 5.5 cm and 20 cm?

4 A rectangle has area 120 cm². Its perimeter is 46 cm. What are the lengths of its sides?

5 Is it possible for a rectangle with area 120 cm² to have a perimeter greater than 120 cm? Explain your answer.

6 If all the side lengths are whole numbers, what is the smallest perimeter that a rectangle with an area of 120 cm² could have?

➡ *Workbook pages 52–53*

Statistics

Use data to answer questions

Think and share

Four pupils are talking about how much writing they do at school.

Naz: I write less than the others.

Claire: I am sure I write more than Tony.

Zara: I write the most.

Tony: I have no idea how many words we write.

How could the pupils decide who is correct?

On Thursday they had five lessons: art, science, maths, geography and English. They each counted how many words they wrote in each lesson and made this table:

	Number of words				
Zara	0	66	10	105	82
Claire	0	81	23	75	86
Naz	3	52	8	99	128
Tony	0	34	11	53	62

Tony says they should have put headings for each subject. Naz says they don't need that information. What do you think?
Does the table tell them who is correct? Explain why or why not.
Can you tell from this table who wrote most words and who wrote fewest words?
What should the pupils do with this information?

1 Draw a table to show how many words each pupil wrote in total.

   a Who wrote the most words?   b Who wrote the fewest words?

   c How many words did they write altogether?

   d Read what the pupils said again. Were any of them correct?

2 Zara drew this **bar chart** to show the data.

   a Why is it difficult to tell how many words each child wrote?

   b How could this bar chart be improved?

   c Redraw the bar chart so you can easily see how many words each pupil wrote.

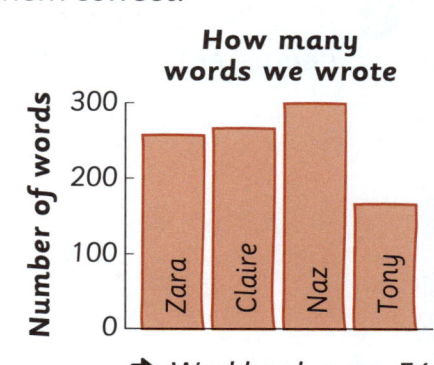

How many words we wrote

▶ *Workbook page 54*

Tables and graphs

A group of pupils recorded how many times they spun the numbers 1 to 6 on a spinner in a **frequency table**.

Score	1	2	3	4	5	6
Frequency	5	8	12	9	3	8

How do you think they kept track of the numbers as they worked? Why?

They drew these graphs to show the data. Which graph do you prefer? Why?

Graph A is a bar chart, Graph B is a **bar-line chart** and Graph C is a **dot plot**.

A
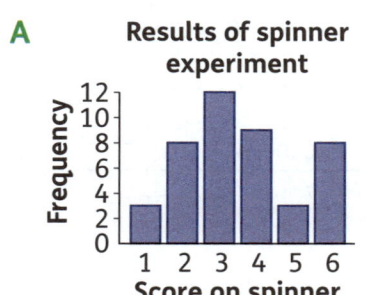

B
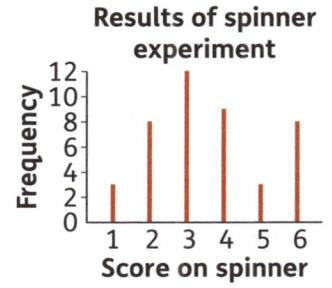

C
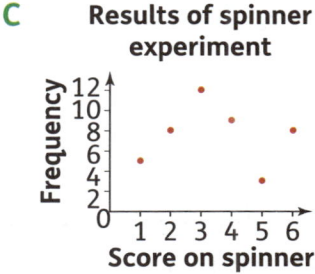

1. Draw a bar chart, a bar-line chart or a dot plot to represent the data in each table. Use a different type of chart for each table.

   a Number of marbles picked up with one hand

Name	Marbles
Paul	14
Alan	11
Katie	10
Chris	13
Ian	16
Nina	12
Bhuddi	15
Alex	12
Monique	13
Laura	14

   b Number of pencils of each colour

Colour	Number
red	25
orange	18
yellow	14
green	22
purple	19
blue	12
pink	28
brown	16

   c Average monthly temperature in Moscow

Month	Temperature (°C)
January	−5
February	−3
March	2
April	11
May	19
June	22
July	24
August	22
September	16
October	8
November	1
December	−3

2. Show your graphs to a partner. Explain why you chose each type and how you decided what scale to use for the data.

➡ *Workbook page 55*

Tables and diagrams

Carroll diagram

	Prime	Not prime
Even	2	4 6 8 10 12 14 16 18 20
Not even	3 5 7 11 13 17 19	1 9 15

Venn diagram

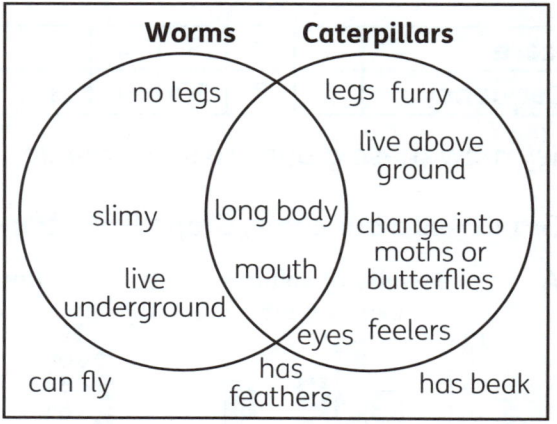

What data do the diagrams show?
How is the data organised?

1. This Venn diagram shows the number of pupils who take art and design classes. How many pupils:

   a do not take art or design classes

   b take both art and design classes

   c take design, but not art

   d take art classes?

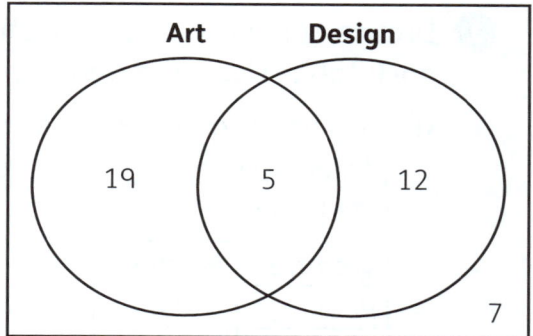

2. This Carroll diagram contains four mistakes.

   a Find them. List them and say what each mistake is.

   b Write another number that could go into each part of the diagram.

	Multiple of 4	Not a multiple of 4
Multiple of 6	12 24 36 40 60	6 18 20 30 42 54
Not a multiple of 6	4 8 16 20 28 32 44 48 52 56	14 25 32 38

Workbook page 56

Line graphs

Line graphs show how something changes over a period of time. Time is usually on the horizontal **axis**.

- Always label both **axes**.
- Mark points on the graph using dots.
- Join the points with lines.
- The graph must have a title.

What does this line graph show? Take turns to ask your partner questions about the graph.

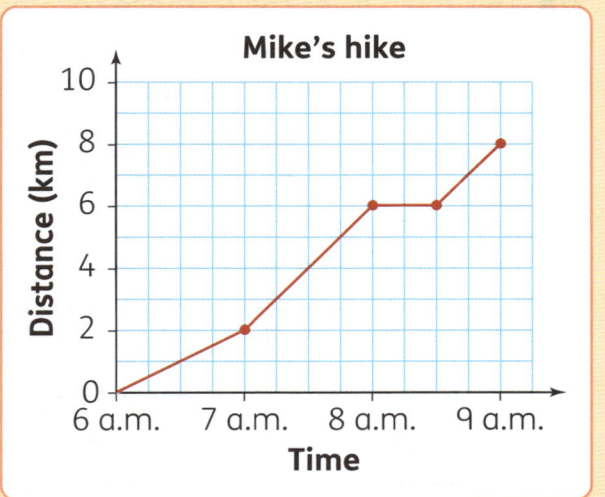

Mike's hike

1 Study this line graph carefully. Then answer the questions.

a What does the graph show?

b What does the horizontal axis show? What units are used?

c What does the vertical axis show? What units are used?

d How many pages did each child read in 5 minutes?

e How long did it take Josh to read 12 pages?

f Who read 16 pages in 4 minutes?

g Which child read faster? How did you decide?

h How long do you think it will take Josh to read 20 pages? Give reasons.

i 'The graph shows that Ani will take 15 minutes to read 50 pages.' Do you agree? Explain your answer.

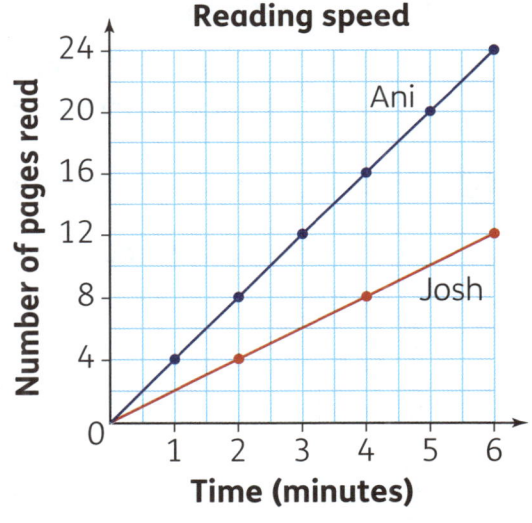

Reading speed

2 How could you collect this data for you and a partner? Discuss your ideas.

➡ *Workbook page 57*

More line graphs

1 This table gives the distance covered by a coastguard boat as it patrolled the coast at a steady speed. Draw a line graph to show this data. Decide on a suitable scale for each axis.

Time (minutes)	Distance (km)
0	0
30	24
60	48
90	72
120	96
150	120
180	144

2 Use your graph to answer these questions.

a How long did it take the coastguard to travel 60 km?

b How far did the boat travel in the first $1\frac{1}{4}$ hours?

c After the boat had travelled 84 km, the coastguard turned round and travelled back to port. How long did the journey back take?

3 This line graph shows how much fuel the boat used in 3 hours.

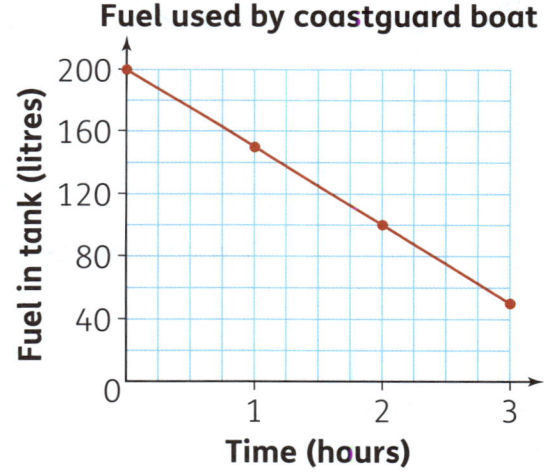

a Why does the line on this graph slope down to the right?

b How much fuel was in the tank to start with?

c How much fuel did the boat use in 3 hours?

d How long did the boat take to use half the fuel that it started with?

➡ *Workbook page 58*

The mode, median and range

The table shows how much six bean plants grew in a week.

Plant	A	B	C	D	E	F
Length grown	10 cm	14 cm	10 cm	11 cm	11 cm	10 cm

The length that appears most often is 10. 10 is the **mode**.
The mode is the value that appears most often in a set of data.

The **median** is the middle value when a set of data is written in order.
In order, the lengths are (in cm): 10, 10, 10, 11, 11, 14.
The middle value is halfway between the third
and fourth value.
The median is 10.5 cm.

middle

10   10   10   11   11   14

The **range** is the difference between the highest and lowest values.
The highest value is 14 cm. The lowest value is 10 cm.
The range is 14 cm − 10 cm = 4 cm.

1. Find the mode, the median and the range of each set of data.

   a 100   95   90   85   80   75   75   75   70   65

   b 4   4   4   8   7   6   6   6   6   7

   c 4   6   2   1   5   6   2   3

 Problem solving

2. Read this information about the heights of 5 pupils.

   The range of our heights is 6 cm.

   The median height is 135 cm.

   The mode of our heights is 134 cm.

   What heights could the pupils be?

    Draw a diagram or make an ordered list.

Frequency tables

Harprit counted the number of letters in 50 words in his library book. He drew this frequency table to organise his results.

Number of letters	Tally	Frequency
1					3
2	ⅲⅲ ||	7
3	ⅲⅲ ⅲⅲ	10
4	ⅲⅲ ⅲⅲ		11
5	ⅲⅲ ||	7
6	ⅲⅲ ||	7
7					3
8			1
11			1

1 How many letters are there in:

   a the longest word

   b the shortest word?

2 What number of letters is the mode?

3 Here are 50 words from another library book.

> John answered, "I am a merchant" and opening his napkin he showed her its contents. Then she exclaimed, "Oh, what beautiful golden things!" And, setting the pails down, she looked at the cups one after another and said, "The king's daughter must see these; she is so pleased with anything ..."

   a Count the number of letters in each word. Make your own frequency table to record the results.

   b How many letters are there in:

     i the longest word

     ii the shortest word?

   c What number of letters is the mode?

4 Which book do you think has the longest words, Harprit's or the other one? Say why you think this.

5 Do your own survey of the number of letters in 50 words. Choose any book for this.

   a Present your data in a frequency table.

   b Write two sentences summarising the data.

Frequency tables with groups

1 Linda surveyed some pupils in her class about the number of minutes they spent on homework one evening. Here are their times.

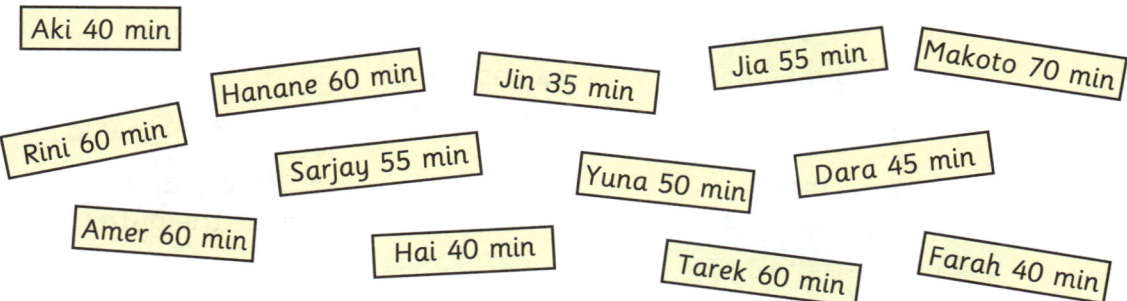

Aki 40 min

Hanane 60 min

Rini 60 min

Jin 35 min

Jia 55 min

Makoto 70 min

Sarjay 55 min

Amer 60 min

Yuna 50 min

Dara 45 min

Hai 40 min

Tarek 60 min

Farah 40 min

Copy this table and complete it.

Time (minutes)	Number of pupils
0–20	
21–40	
41–60	
61–80	

2 Each pupil in a small school grew a sunflower.

Here are the heights (in centimetres) after 4 months:

98, 102, 155, 220, 253, 94, 149, 190, 192, 234, 251, 112, 165, 249, 252, 132, 171, 230, 250, 127, 186, 201, 105, 175, 111, 173, 207, 150, 185, 200, 206, 119, 162, 198, 204, 142, 176, 154, 205, 189, 173, 144, 200, 164, 190, 100, 178, 151, 112, 156

Copy and complete the frequency table.

Height (cm)	Frequency
50–100	
101–150	
151–200	
201–250	
251–300	

➡ *Workbook page 59*

Do your own investigation

These are the steps in a statistical investigation.

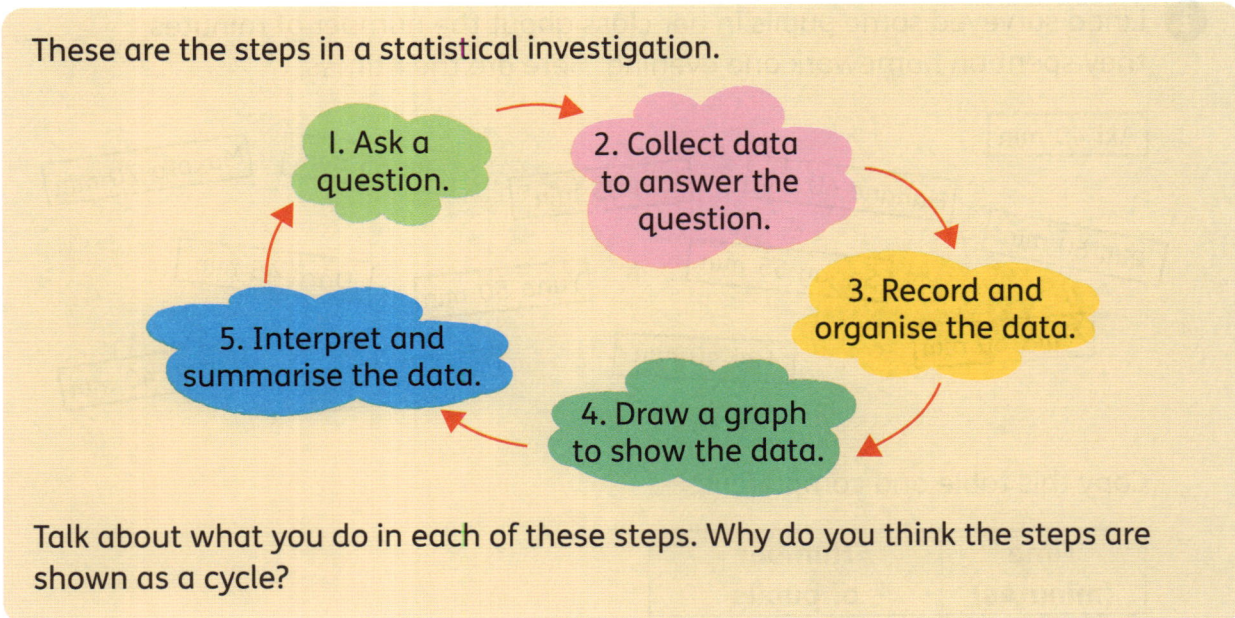

1. Ask a question.

2. Collect data to answer the question.

3. Record and organise the data.

4. Draw a graph to show the data.

5. Interpret and summarise the data.

Talk about what you do in each of these steps. Why do you think the steps are shown as a cycle?

1 Look at this graph carefully. Discuss what it shows.

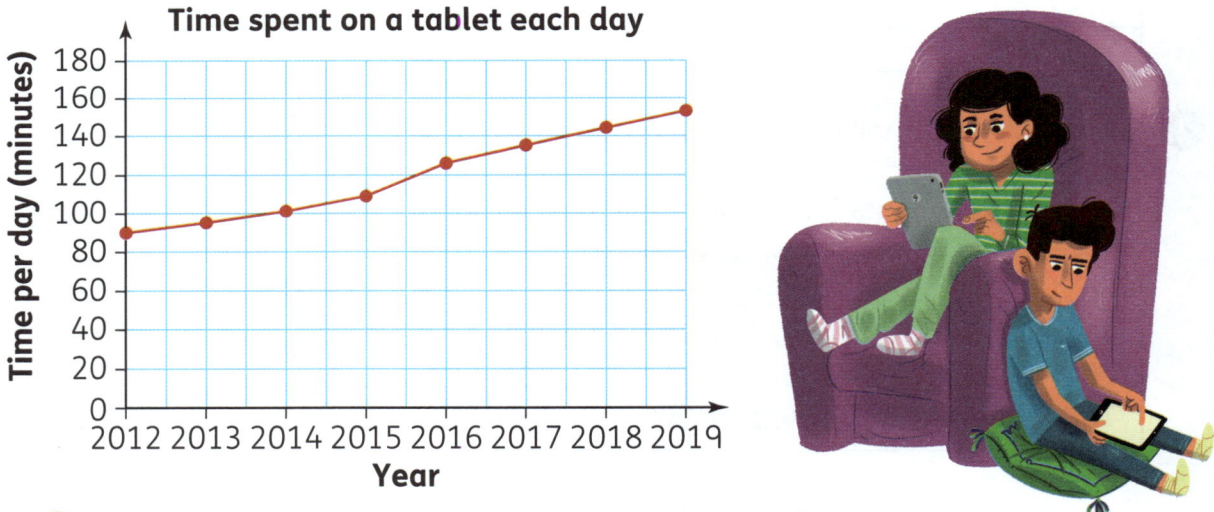

Time spent on a tablet each day

Time per day (minutes) — Year

2 Work in pairs to do your own statistical investigation.

a What questions could you ask to find out how long pupils at your school spend on a tablet each day?

b Choose one question. Plan how to collect and organise the data you need to answer the question.

c Collect and organise the data.

d Draw at least one graph to show your data.

e Write a paragraph summarising what you found out.

Timetables

This timetable shows the arrival and departure times of a long-distance bus.

Bus station	Arrival time	Departure time
Nairobi	—	08:30
Nakuru	11:45	12:20
Eldoret	14:45	15:30
Meru	16:25	16:35
Thika	18:30	19:00
Machakos	21:05	21:15
Nairobi	22:00	—

You can use data from the timetable to work out how long a journey will take.

How long does the journey from Meru to Thika take?
Draw a number line and count in steps to work this out.

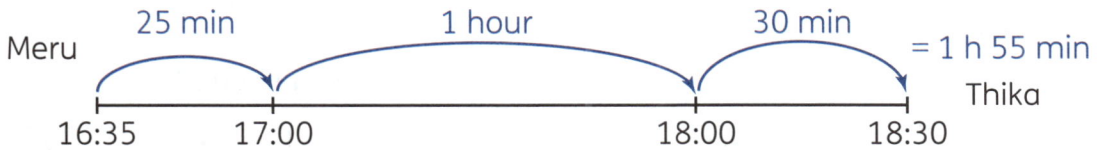

Why is it difficult to use a calculator for time calculations?

1. Answer these questions about the bus timetable.
   a When does the bus depart from Nairobi?
   b At what time does the bus arrive in Eldoret?
   c For how long does the bus stop in Meru?
   d Where does the bus stop for the longest time? How long is this stop?
   e When does the bus get back to Nairobi?

2. Calculate the journey time (including stops) from:
   a Nairobi to Nakuru   b Nakuru to Eldoret   c Nairobi to Nairobi

3. How much time does the bus spend travelling between 08:30 and 21:55?

4. On Monday, the bus left Thika on time and arrived in Machakos 15 minutes early. What was its arrival time?

5. On Tuesday, the bus got a puncture between Eldoret and Meru. This made the bus 45 minutes late. At what time did the bus arrive in Meru?

Revisit fractions

Think and share

Tell your group five things that you remember about fractions.

This rectangle has been divided into parts.

Has it been divided into quarters? Give a reason for your answer.
What fraction of the shape is each part?

Is it correct to say that the whole rectangle represents $\frac{16}{16}$? Give a reason for your answer.

1 For each diagram, write the fraction:

   a that is shaded   **b** that is unshaded

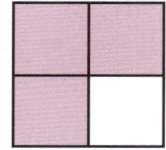  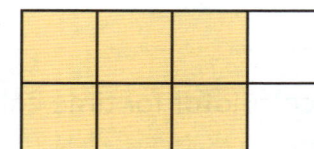 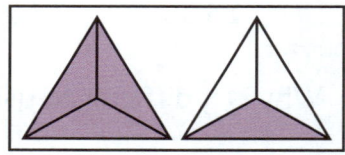

2 Write the fraction shown by the arrow on each number line. Count how many parts the line is divided into to help you.

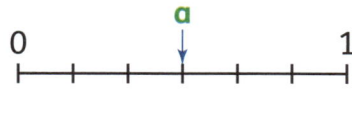  

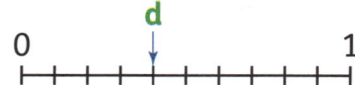  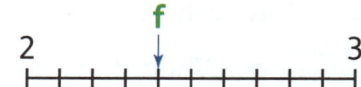

3 How many fractions can you write with a denominator of 6 and a value of less than 1?

4 Look at the circle.

   a How many parts is it divided into?

   b What fraction is shown by each shaded section?

   c Is the sum of the unshaded sections greater or less than $\frac{1}{4}$ of the circle?

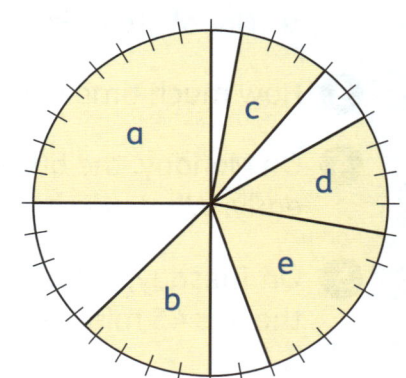

Equivalent fractions

Equivalent fractions have the same value.
Equivalent fraction charts or number lines help you to find equivalent fractions.

```
|_____|_____|        |____|____|____|____|        |__|__|__|__|__|__|__|__|
0        1/2      1        0    1/4  2/4  3/4  1         0  1/8 2/8 3/8 4/8 5/8 6/8 7/8 1
```

The number lines show that: $\frac{1}{2} = \frac{2}{4} = \frac{4}{8}$ $\frac{1}{4} = \frac{2}{8}$ $\frac{3}{4} = \frac{6}{8}$

1 whole		
$\frac{1}{3}$	$\frac{1}{3}$	$\frac{1}{3}$

$\frac{1}{6}$	$\frac{1}{6}$	$\frac{1}{6}$	$\frac{1}{6}$	$\frac{1}{6}$	$\frac{1}{6}$

$\frac{1}{12}$	$\frac{1}{12}$	$\frac{1}{12}$	$\frac{1}{12}$	$\frac{1}{12}$	$\frac{1}{12}$	$\frac{1}{12}$	$\frac{1}{12}$	$\frac{1}{12}$	$\frac{1}{12}$	$\frac{1}{12}$	$\frac{1}{12}$

$$\overset{\times 2}{\frown}\ \overset{\times 3}{\frown}\qquad \overset{\div 7}{\frown}\qquad \overset{\div 2}{\frown}$$

$$\frac{4}{5} = \frac{8}{10} = \frac{24}{30}\qquad \frac{7}{14} = \frac{1}{2}\qquad \frac{16}{30} = \frac{8}{15}$$

$$\underset{\times 2}{\smile}\ \underset{\times 3}{\smile}\qquad \underset{\div 7}{\smile}\qquad \underset{\div 2}{\smile}$$

The chart shows that:

$\frac{1}{3} = \frac{2}{6} = \frac{4}{12}$ $\frac{2}{3} = \frac{4}{6} = \frac{8}{12}$

$\frac{1}{6} = \frac{2}{12}$ and $\frac{5}{6} = \frac{10}{12}$

1 Work with a partner to work out the missing **numerator** in each pair.

a $\frac{2}{10} = \frac{\square}{5}$ b $\frac{1}{4} = \frac{\square}{12}$ c $\frac{6}{8} = \frac{\square}{4}$ d $\frac{9}{27} = \frac{\square}{3}$

e $\frac{1}{3} = \frac{\square}{21}$ f $\frac{15}{20} = \frac{\square}{4}$ g $\frac{16}{24} = \frac{\square}{12}$ h $\frac{16}{24} = \frac{\square}{6}$

> To find an equivalent fraction, multiply or divide the numerator and denominator by the same number.

2 Write three equivalent fractions for each of these.

a $\frac{1}{6}$ b $\frac{3}{6}$ c $\frac{2}{3}$ d $\frac{4}{6}$ e $\frac{8}{10}$ f $\frac{2}{8}$

g $\frac{2}{5}$ h $\frac{6}{8}$ i $\frac{1}{5}$ j $\frac{6}{10}$ k $\frac{3}{6}$ l $\frac{3}{5}$

3 Write the next three equivalent fractions in each sequence.

a $\frac{1}{2}, \frac{2}{4}, \frac{3}{6}, \ldots$ b $\frac{1}{5}, \frac{2}{10}, \frac{3}{15}, \ldots$ c $\frac{3}{8}, \frac{6}{16}, \frac{9}{24}, \ldots$

4 $\frac{3}{4}$ and $\frac{9}{12}$ are equivalent fractions.

Draw diagrams to show that even though these fractions are equivalent, they may not represent the same amount.

➡ *Workbook pages 60–61*

Improper fractions and mixed numbers

The diagram shows 5 halves.

We can write this as an **improper fraction**: $\frac{5}{2}$.

If we combine 5 halves to make wholes, we get 2 wholes and 1 half.

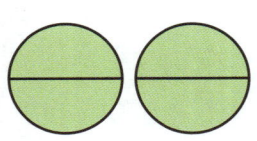

We can write this as a **mixed number**: $2\frac{1}{2}$

Read through these examples to see two ways of converting $\frac{7}{6}$ to a mixed number.

Using diagrams

$\frac{7}{6} =$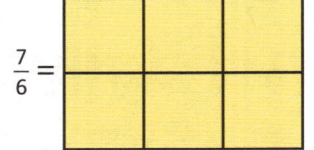

This is $\frac{6}{6} + \frac{1}{6} = 1 + \frac{1}{6} = 1\frac{1}{6}$

Using division

$\frac{7}{6}$ means $7 \div 6$

7 divided by 6 is 1 with 1 left over.
The 1 that is left over is 1 of 6 pieces or $\frac{1}{6}$

$\frac{7}{6} = 1\frac{1}{6}$

1 Write these improper fractions as mixed numbers.

a $\frac{3}{2}$ b $\frac{4}{3}$ c $\frac{5}{4}$ d $\frac{7}{4}$ e $\frac{9}{4}$ f $\frac{15}{7}$

g $\frac{12}{5}$ h $\frac{13}{3}$ i $\frac{19}{4}$ j $\frac{7}{3}$ k $\frac{11}{6}$ l $\frac{25}{6}$

2 Which is greater in each pair?

a $5\frac{1}{2}$ or $\frac{7}{2}$ b $2\frac{3}{4}$ or $\frac{7}{4}$ c $1\frac{3}{5}$ or $\frac{9}{5}$ d $3\frac{2}{7}$ or $\frac{17}{7}$ e $\frac{4}{3}$ or $1\frac{2}{3}$ f $\frac{14}{4}$ or $3\frac{3}{4}$

g $7\frac{1}{2}$ or $\frac{20}{2}$ h $\frac{19}{8}$ or $2\frac{3}{8}$ i $12\frac{1}{3}$ or $\frac{40}{3}$ j $\frac{19}{5}$ or $3\frac{3}{5}$ k $7\frac{2}{7}$ or $\frac{48}{7}$ l $\frac{50}{8}$ or $6\frac{7}{8}$

3 Write each set in order, from smallest to greatest.

a $1\frac{1}{2}$ $\frac{7}{3}$ $\frac{9}{5}$ $2\frac{1}{4}$ $\frac{15}{6}$

b $\frac{8}{2}$ $\frac{3}{4}$ $\frac{12}{5}$ $\frac{19}{6}$ $\frac{21}{5}$

c $\frac{12}{5}$ $\frac{7}{3}$ $\frac{3}{2}$ $\frac{9}{4}$ $\frac{12}{8}$

▶ *Workbook page 62*

Compare and order fractions

We can use equivalent fractions to compare fractions with different denominators.

Which is greater: $\frac{2}{3}$ or $\frac{3}{4}$?

$$\frac{2}{3} \overset{\times 4}{\underset{\times 4}{=}} \frac{8}{12} \qquad \frac{3}{4} \overset{\times 3}{\underset{\times 3}{=}} \frac{9}{12}$$

$\frac{9}{12} > \frac{8}{12}$ so $\frac{3}{4} > \frac{2}{3}$

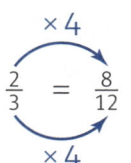 Find a common multiple of 3 and 4 to use as the denominator.

Which is less: $2\frac{1}{3}$ or $\frac{8}{3}$?

$2\frac{1}{3} = \frac{3}{3} + \frac{3}{3} + \frac{1}{3} = \frac{7}{3}$

$\frac{7}{3} < \frac{8}{3}$ so $2\frac{1}{3} < \frac{8}{3}$

To compare mixed numbers and improper fractions, write them both as mixed numbers or both as improper fractions. Sometimes you might need to find an equivalent improper fraction.

Compare $1\frac{3}{4}$ and $\frac{19}{12}$.

$1\frac{3}{4} = \frac{4}{4} + \frac{3}{4} = \frac{7}{4}$

Convert this to an equivalent fraction with denominator 12.

$$\frac{7}{4} \overset{\times 3}{\underset{\times 3}{=}} \frac{21}{12}$$

$\frac{21}{12} > \frac{19}{12}$ so $1\frac{3}{4} > \frac{19}{12}$

---

1 Compare each set of fractions using the < or > sign.

a $\frac{3}{8}$ and $\frac{1}{4}$ b $\frac{1}{3}$ and $\frac{4}{10}$ c $\frac{2}{3}$ and $\frac{3}{5}$ d $\frac{19}{8}$ and $\frac{12}{4}$

e $\frac{3}{4}$ and $\frac{7}{10}$ f $2\frac{1}{2}$ and $\frac{5}{3}$ g $\frac{11}{4}$ and $3\frac{2}{3}$ h $1\frac{7}{10}$ and $\frac{9}{5}$

 Draw a diagram if you need to.

Problem solving

2 Mara and Alisa had the same amount of candy at the start of the day. At the end of the day, Mara had $\frac{11}{15}$ left and Alisa had $\frac{4}{5}$ left. Who had the most candy left?

3 The amount of liquid in five containers is given here.

a Write the amounts in order, from most liquid to least liquid.

b Which containers are more than half full?

$\frac{3}{8}\ell$ $\frac{4}{15}\ell$ $\frac{2}{3}\ell$ $\frac{3}{5}\ell$ $\frac{9}{12}\ell$

➡ *Workbook page 63*

Add and subtract fractions

This strip is divided into tenths.

$$\frac{2}{10} + \frac{4}{10} = \frac{6}{10}$$ $$\frac{3}{10} + \frac{1}{10} = \frac{4}{10}$$

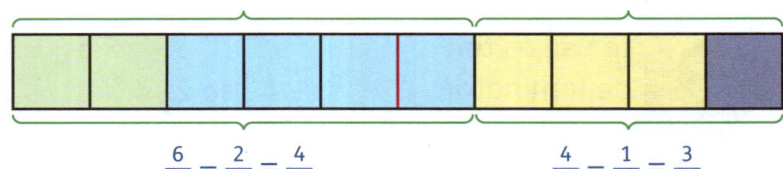

$$\frac{6}{10} - \frac{2}{10} = \frac{4}{10}$$ $$\frac{4}{10} - \frac{1}{10} = \frac{3}{10}$$

- Why don't we add or subtract the denominators?

- Can you use the diagram to work out $1 - \frac{4}{10}$?

- What is $\frac{9}{10} - \frac{2}{10}$?

Look at these examples.

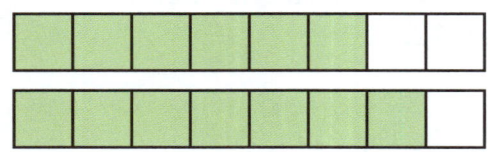

$$\frac{6}{8} + \frac{7}{8} = \frac{13}{8}$$

$$\frac{3}{4} + \frac{3}{4} = \frac{6}{4}$$

1 Work with a partner to do these calculations. Draw diagrams if you need to.

a $1 - \frac{1}{2}$ b $3\frac{7}{9} + 1\frac{1}{9}$ c $\frac{4}{5} + \frac{4}{5}$ d $2\frac{3}{5} - 1\frac{2}{5}$

2 Complete these calculations.

a $\frac{3}{5} + \frac{1}{5}$ b $\frac{9}{11} + \frac{1}{11}$ c $\frac{4}{9} + \frac{3}{9}$ d $\frac{11}{20} + \frac{5}{20}$

e $\frac{9}{11} - \frac{3}{11}$ f $\frac{9}{12} - \frac{6}{12}$ g $1\frac{1}{20} - 1\frac{0}{20}$ h $\frac{7}{8} - \frac{5}{8}$

> Think carefully and draw a diagram if you need to.

3 Calculate. Give your answers as mixed numbers where appropriate.

a $2\frac{3}{8} - \frac{7}{8}$ b $3\frac{2}{7} - 1\frac{4}{7}$ c $3\frac{2}{5} - 1\frac{4}{5}$ d $4\frac{1}{15} - 1\frac{9}{15}$

e $1\frac{4}{7} + \frac{4}{7}$ f $1\frac{9}{10} + \frac{8}{10}$ g $2\frac{3}{7} + 2\frac{6}{7}$ h $1\frac{2}{3} + 1\frac{2}{3}$

 Problem solving

4 Zara bought $2\frac{1}{4}$ kg of onions, $1\frac{1}{4}$ kg of garlic and $2\frac{3}{4}$ kg of tomatoes. What is the total mass of her shopping?

More adding and subtracting

Tell your partner how you would work out the answers to these problems.

How many metres of tape are on the three rolls altogether?

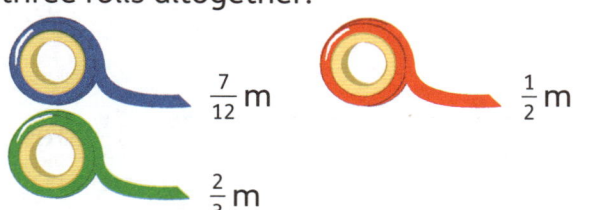

$\frac{7}{12}$ m $\frac{1}{2}$ m

$\frac{2}{3}$ m

How much heavier is the second bag of fruit?

$1\frac{9}{10}$ kg $2\frac{3}{5}$ kg

When adding or subtracting fractions with different denominators, use equivalent fractions to change to fractions with the same denominators.

$\frac{3}{5} + \frac{7}{10}$

$\frac{6}{10} + \frac{7}{10} = \frac{13}{10}$

$\frac{3}{5} = \frac{6}{10}$

13 tenths is 1 whole and $\frac{3}{10}$, or $1\frac{3}{10}$.

$1\frac{5}{12} + 2\frac{1}{4}$

$1 + 2 = 3$

$\frac{5}{12} + \frac{1}{4} = \frac{5}{12} + \frac{3}{12} = \frac{8}{12}$

$3 + \frac{8}{12} = 3\frac{8}{12} = 3\frac{2}{3}$

$2\frac{1}{3} - 1\frac{5}{6}$

$2\frac{1}{3} = \frac{3}{3} + \frac{3}{3} + \frac{1}{3} = \frac{7}{3}$

$1\frac{5}{6} = \frac{6}{6} + \frac{5}{6} = \frac{11}{6}$

$\frac{7}{3} - \frac{11}{6} = \frac{14}{6} - \frac{11}{6} = \frac{3}{6} = \frac{1}{2}$

1 Use equivalent fractions to help you to calculate these.

a $\frac{4}{5} + \frac{5}{20}$ b $\frac{7}{4} + \frac{1}{8}$ c $\frac{7}{12} + \frac{1}{4}$ d $\frac{1}{2} + \frac{3}{10}$

e $\frac{7}{8} - \frac{1}{4}$ f $\frac{3}{4} - \frac{1}{8}$ g $\frac{5}{18} - \frac{1}{9}$ h $\frac{15}{10} - \frac{12}{30}$

2 Add. Show your working. Give your answers as mixed numbers in their simplest form.

a $1\frac{3}{4} + 2\frac{1}{8}$ b $1\frac{1}{4} + 3\frac{1}{12}$ c $3\frac{7}{8} + 1\frac{1}{4}$ d $3\frac{2}{3} + 1\frac{1}{6}$

e $2\frac{3}{10} + 1\frac{1}{2}$ f $2\frac{7}{20} + 1\frac{1}{4}$ g $4\frac{3}{5} + 1\frac{6}{10}$ h $3\frac{1}{2} + 1\frac{7}{8}$

3 Subtract. Show your working. Give your answers as mixed numbers in their simplest form.

a $4\frac{1}{2} - 1\frac{3}{10}$ b $3\frac{2}{3} - 1\frac{1}{6}$ c $5\frac{1}{2} - 1\frac{5}{12}$ d $4\frac{3}{8} - 1\frac{1}{4}$

e $3\frac{1}{2} - 1\frac{6}{8}$ f $4\frac{3}{4} - 1\frac{1}{2}$ g $4\frac{3}{12} - 1\frac{2}{3}$ h $\frac{11}{4} - 1\frac{3}{8}$

Workbook page 64

Multiply fractions by whole numbers

> When you work with real measurements, give the answer as a mixed number rather than an improper fraction.

Every day Maryam drinks $\frac{2}{3}$ of a litre of water. How much water does she drink in 5 days?

Here are some ways to work this out.

Use a diagram:

$\frac{10}{3} = 3\frac{1}{3}$ litres

You can also use multiplication:

$5 \times \frac{2}{3}$ Think 5 times 2 thirds = 10 thirds.

$\frac{5}{1} \times \frac{2}{3} = \frac{10}{3}$ Multiply denominator by denominator and numerator by numerator.

$\frac{10}{3} = 3\frac{1}{3}$ litres

Use repeated addition:

$\frac{2}{3} + \frac{2}{3} + \frac{2}{3} + \frac{2}{3} + \frac{2}{3} = \frac{10}{3}$ litres
$= 3\frac{1}{3}$ litres

Any whole number can be written as a fraction with denominator 1.

1 Multiply. Try to use a written method. If the answer is an improper fraction, write it as a mixed number too.

> Draw a diagram if you need to.

a $2 \times \frac{3}{5}$ b $2 \times \frac{1}{7}$ c $3 \times \frac{3}{5}$ d $7 \times \frac{3}{10}$

e $5 \times \frac{4}{7}$ f $12 \times \frac{3}{4}$ g $10 \times \frac{2}{3}$ h $16 \times \frac{1}{5}$

i $11 \times \frac{3}{7}$ j $25 \times \frac{1}{2}$ k $14 \times \frac{1}{4}$ l $5 \times \frac{7}{8}$

2 Copy each number sentence. Fill in the missing numbers.

a $2 \times \frac{\square}{7} = \frac{6}{7}$ b $3 \times \frac{9}{\square} = \frac{9}{10}$ c $5 \times \frac{5}{\square} = \frac{5}{9}$

d $2 \times \frac{\square}{\square} = \frac{2}{7}$ e $2 \times \frac{\square}{16} = \frac{5}{8}$ f $3 \times \frac{\square}{9} = \frac{2}{3}$

g $3 \times \frac{\square}{12} = \frac{1}{4}$ h $5 \times \frac{\square}{20} = \frac{3}{4}$ i $3 \times \frac{\square}{\square} = \frac{7}{10}$

Problem solving

3 Choose any three calculations from question 1. Make up a word problem to match each one.

Multiply with mixed numbers

Lisa is baking 4 loaves of bread. She needs $1\frac{1}{3}$ cups of buttermilk for each loaf. How can she work out how much buttermilk she needs altogether?

Here are some ways to work this out.

Draw and count or add.

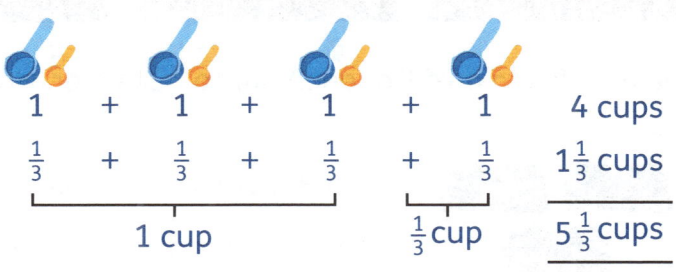

| 1 | $+$ | 1 | $+$ | 1 | $+$ | 1 | 4 cups |
| $\frac{1}{3}$ | $+$ | $\frac{1}{3}$ | $+$ | $\frac{1}{3}$ | $+$ | $\frac{1}{3}$ | $1\frac{1}{3}$ cups |

$\underbrace{\qquad\qquad}_{\text{1 cup}}$ $\underbrace{\quad}_{\frac{1}{3}\text{ cup}}$ $5\frac{1}{3}$ cups

Multiply the whole number and the fraction separately.

$$1\frac{1}{3} \times 4 = 1 \times 4 + \frac{1}{3} \times 4 = 4 + \frac{4}{3} = 4 + 1\frac{1}{3} = 5\frac{1}{3} \text{ cups}$$

Change the mixed number to an improper fraction and multiply.

$$1\frac{1}{3} = \frac{4}{3} \qquad \frac{4}{3} \times \frac{4}{1} = \frac{16}{3} = 5\frac{1}{3} \text{ cups}$$

Discuss each method. Which written method seems easiest to you? Why?

1 Use a written method to do these multiplications. Draw a rough sketch to help you if you need it. Give the answers as mixed numbers.

a $2 \times 2\frac{3}{5}$ b $3 \times 4\frac{1}{2}$ c $4 \times 1\frac{3}{8}$ d $4 \times 2\frac{2}{3}$

e $3 \times 1\frac{4}{9}$ f $4 \times 2\frac{4}{5}$ g $3 \times 4\frac{5}{8}$ h $3 \times 2\frac{9}{10}$

 Problem solving

2 Ahmed ran $4\frac{3}{4}$ km every day for a week.

a How far did he run in total?

b Naresh ran $1\frac{1}{4}$ km less than Ahmed each day. How many kilometres did Naresh run in a week?

3 You need $1\frac{1}{3}$ metres of fabric to make a pair of pyjama trousers.

a How much fabric will a factory need to make 70 pairs of trousers?

b The fabric comes in a roll of 100 metres. How much will be left after making 70 pairs of trousers?

Fractions of an amount

We can use multiplication to work out fractions of things we can count or measure.

Cici has 3 bars of chocolate.

She decides to give $\frac{2}{5}$ of each bar to her sister. How much is $\frac{2}{5}$ of 3 bars?

$\frac{2}{5}$ of 3 means the same as $\frac{2}{5} \times 3$.

$\frac{2}{5} \times 3 = \frac{6}{5}$, which is the same as 1 whole bar and $\frac{1}{5}$ of a bar.

So, $\frac{2}{5}$ of 3 = $1\frac{1}{5}$

What is $\frac{3}{10}$ of 2 metres? Give your answer as a decimal.

$\frac{3}{10}$ of 2 is the same as $\frac{3}{10} \times 2$. $\qquad$ $\frac{3}{10} \times \frac{2}{1} = \frac{6}{10}$

$\frac{6}{10}$ of a metre = 0.6 metres. $\qquad$ So, $\frac{3}{10}$ of 2 metres is 0.6 metres.

---

1 Calculate these amounts.

    a $\frac{1}{2}$ of £2.50 $\qquad$ **b** $\frac{1}{4}$ of 1 metre $\qquad$ **c** $\frac{3}{10}$ of 2 litres

    d $\frac{3}{8}$ of 32 kilograms $\qquad$ **e** $\frac{3}{4}$ of £4.80 $\qquad$ **f** $\frac{7}{8}$ of a kilometre

    g $\frac{6}{10}$ of 80 pence $\qquad$ **h** $\frac{2}{3}$ of £39 $\qquad$ **i** $\frac{5}{6}$ of 480 grams

    j $\frac{5}{8}$ of 200 m $\qquad$ **k** $\frac{3}{4}$ of 5 hours $\qquad$ **l** $\frac{9}{10}$ of 2000 metres

---

💡 Problem solving

2 A shopkeeper sold $\frac{3}{5}$ of a box of 150 mangoes. How many mangoes were left?

3 Josh had £47.75. He spent $\frac{1}{5}$ on an app and $\frac{2}{5}$ on a game. How much money did he have left after that?

4 A rainwater tank that holds 5000 litres is full at the start of April. By the end of the month, $\frac{5}{8}$ of the water has been used. How much water is left in the tank?

➡ *Workbook page 65*

More work with fractions

You can draw diagrams to help you solve problems involving fractions.

Work in pairs. Read the problems. How do the diagrams help you see how to solve each problem? Work out the solution to each problem.

A chef pours some juice from a 1 litre jug. She estimates that she has poured out $\frac{6}{10}$ of a litre, but when she measures, she has actually poured $\frac{1}{5}$ of a litre more than she estimated. How much has she poured?

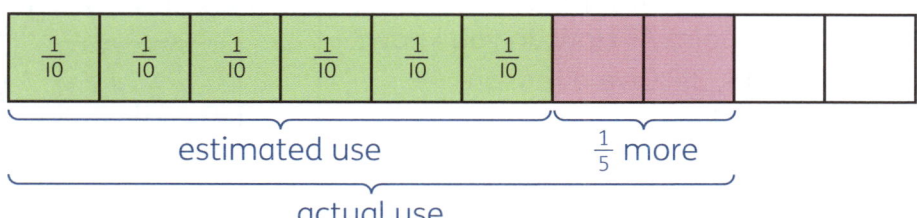

A tourist spends $\frac{3}{4}$ of £300 to rent a car. He spends $\frac{1}{3}$ of what he has left on petrol. How much money does he have left?

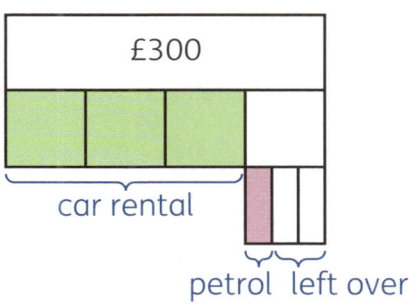

£300

car rental

petrol left over

Problem solving

1. In a 100 m² plot, $\frac{1}{8}$ of the plot is used to grow spinach, $\frac{1}{2}$ is used to grow tomatoes, $\frac{1}{3}$ is used to grow pumpkins and the rest is used to grow onions. What area of land is used for onions?

Draw a diagram if it helps.

2. To raise money for charity, Jo walked $12\frac{1}{3}$ km and Lee walked $14\frac{3}{4}$ km. How much further did Lee walk than Jo?

3. Paula walks $3\frac{1}{2}$ km to school every day. She stops $\frac{3}{4}$ of the way to pick up her friend Lin. How far do Paula and Lin still have to walk?

4. There is a bowl of plums on a table. Tarek eats $\frac{1}{4}$ of them. Then Zara eats $\frac{1}{3}$ of those that are left. After that, Ria eats $\frac{1}{2}$ of those still left. What fraction of the plums is left after that?

Position, direction and movement

Coordinates on a grid

💭 Think and share

Nala drew these two shapes on a **coordinate grid** but she didn't label the axes.

Which is the x-axis and which is the y-axis?
What are the **coordinates** of point A? How do you know?
Is it possible that point B is (3, 3)? Give a reason.
What could the coordinates of point C be? Why?

Rohit says: 'The coordinates of D must be 4 and a number greater than 4.' Explain why this is correct.

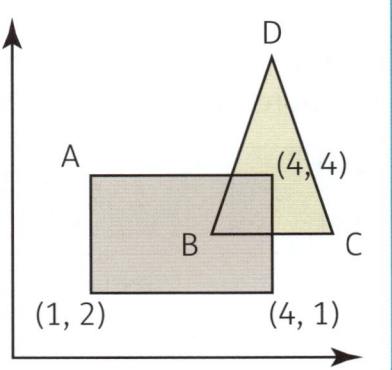

Use this map on a coordinate grid to answer the questions.

1. What is at these coordinates?

   a (5, 1)   b (3, 6)

2. Give the coordinates of these places.

   a Ruby Cavern

   b Hall of Wells

3. Simone is at the entrance at (5, 7). She moves 2 squares down and 2 squares to the left. She wants to get from here to the Dragon Pit. How many squares down and to the right should she move?

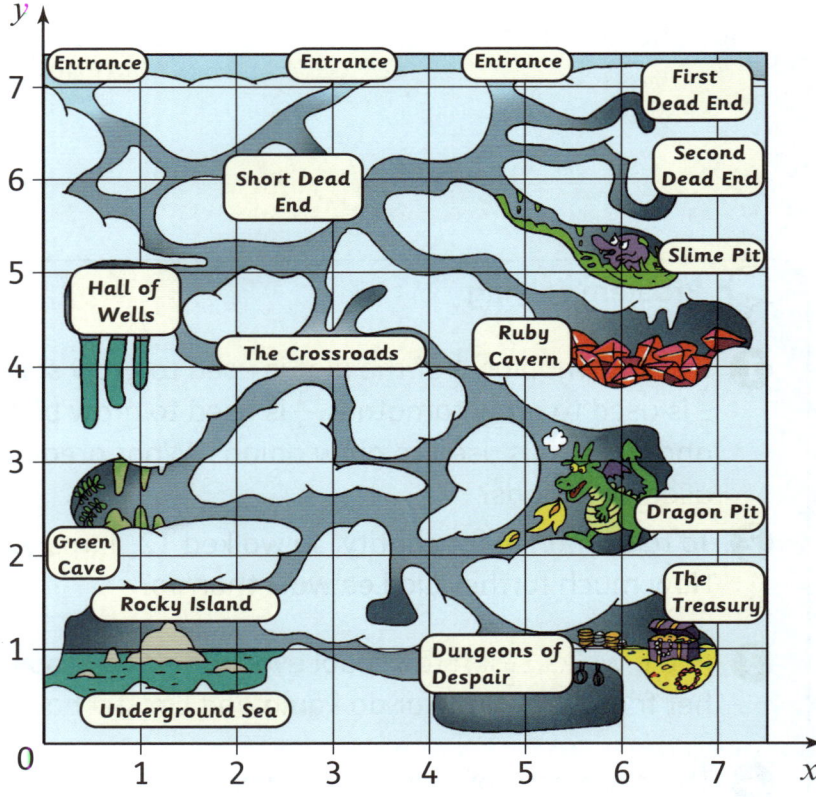

➡ *Workbook page 66*

Shapes on a grid

We use coordinates to plot points on a grid. We can join the points to form shapes.

What polygon have you drawn? How do you know this?

Plot the points A (3, 6), B (7, 6), C (7, 2) and D (3, 2) on a grid.

Join A to B, B to C, C to D and D to A.

ABCD is a square because the sides meet at right angles and all the sides are four grid squares long.

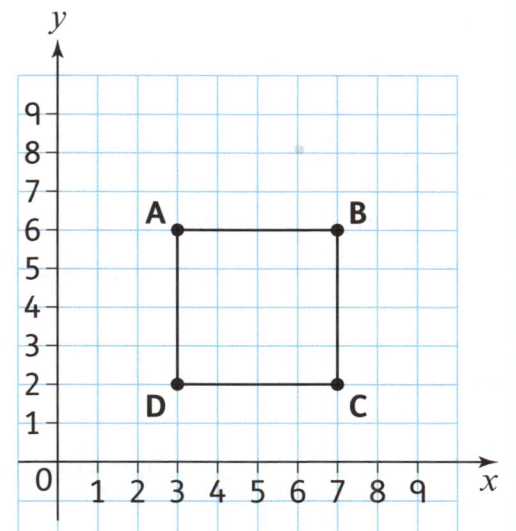

1 Answer these questions about shapes PQRS and XYZ on the grid.

a What are the coordinates of the vertices of PQRS?

b Explain why PQRS is a quadrilateral but not a trapezium.

c How could you make PQRS into a trapezium by moving one point to another position?

d What are the coordinates of the vertices of XYZ?

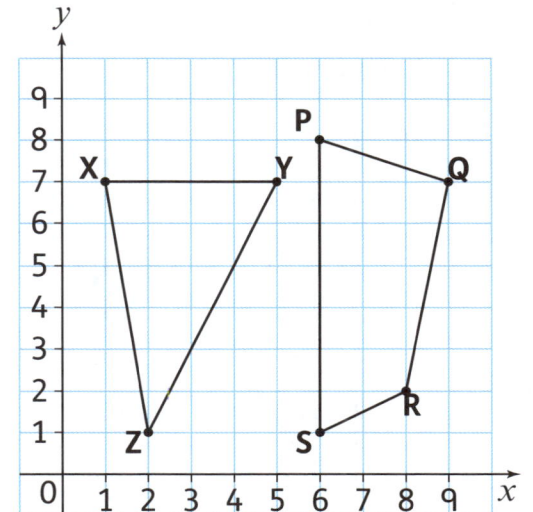

e How would you classify triangle XYZ? Why?

f If point Z moves to (1, 1), what type of triangle will XYZ be? Why?

2 The points A, B, C and D are plotted and joined in order.

A (2, 4), B (1, 0), C (7, 0), D (8, 4)

What shape is ABCD?

3 Write a set of coordinates for drawing a hexagon with at least one of its vertices on the y-axis.

Reflect shapes on a grid

Layla held up a triangle against a mirror.

The **reflection** showed the same triangle pointing in the opposite direction.

In maths, a reflection is a change in the position of a shape. It is also known as a 'flip', because the shape looks as if it has been flipped over to the opposite direction.

The dotted line in the diagram is called the **mirror line**.

Matching points on a shape and its reflection are exactly the same distance from the mirror line.

1 Which shape (A or B) is a reflection of the shape without a letter on each grid? Show your partner where the mirror line is for the shape you have chosen.

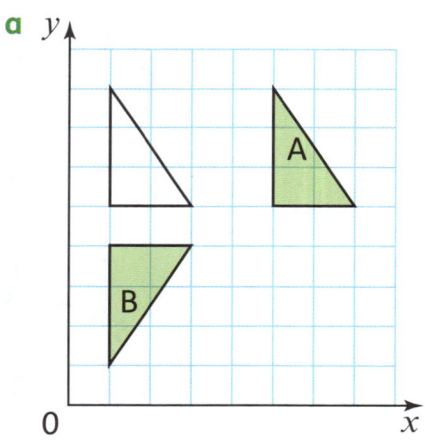

a

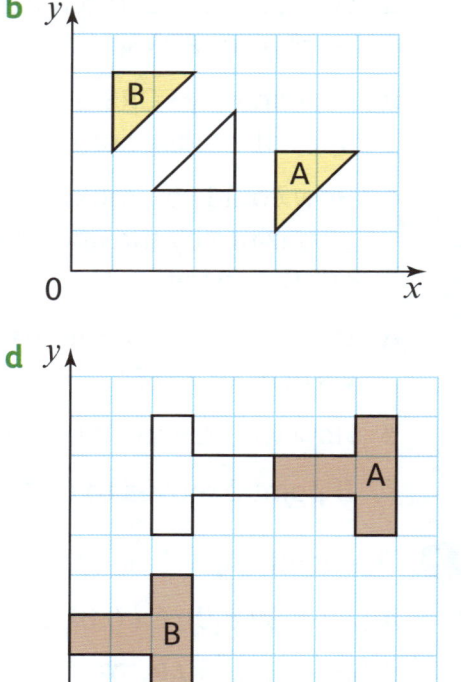

b

c

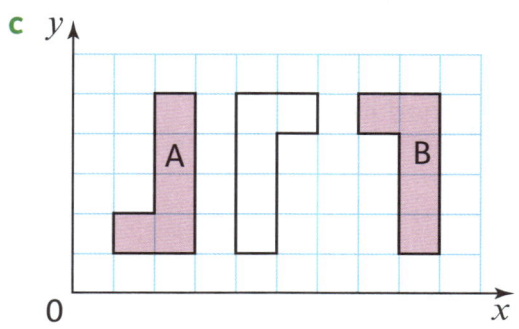

d

➡ *Workbook page 67*

Translate shapes on a grid

A **translation** is also called a slide. You can slide (or translate) a shape right, left, up or down.

Triangle ABC has been translated 9 squares to the right.

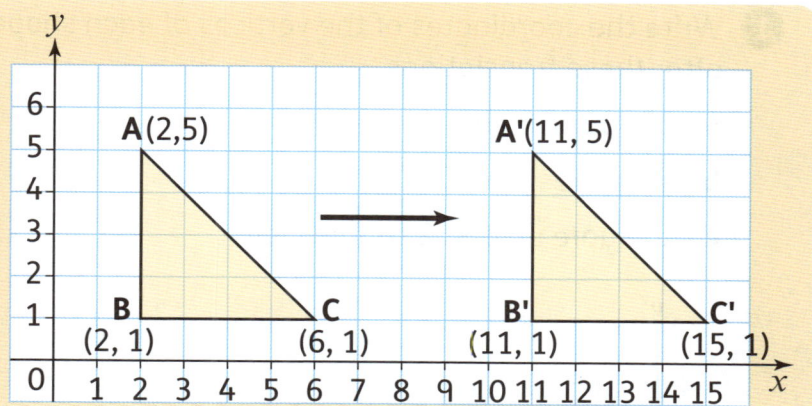

1 Say how many squares right, left, up or down each shape has been translated.

a

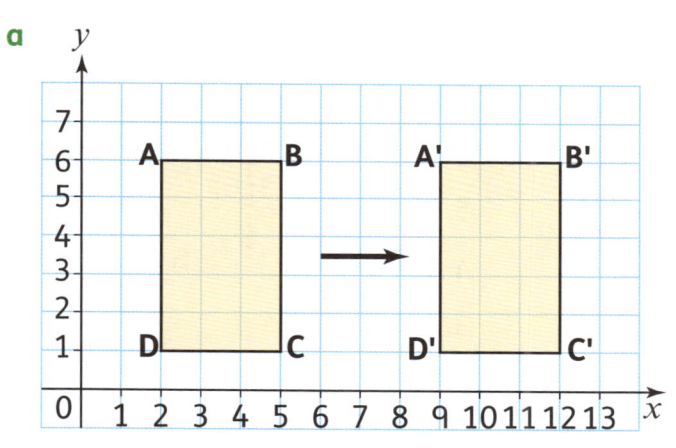

b

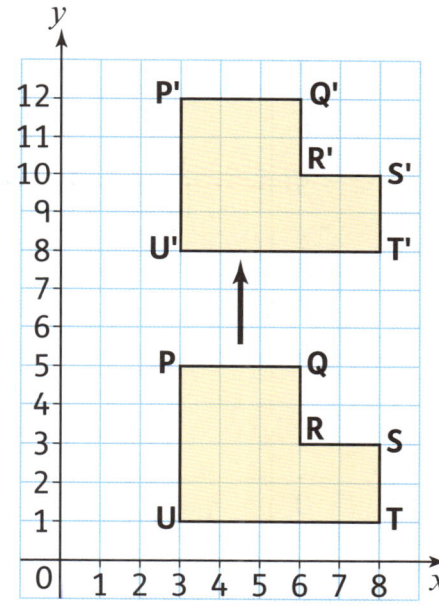

c

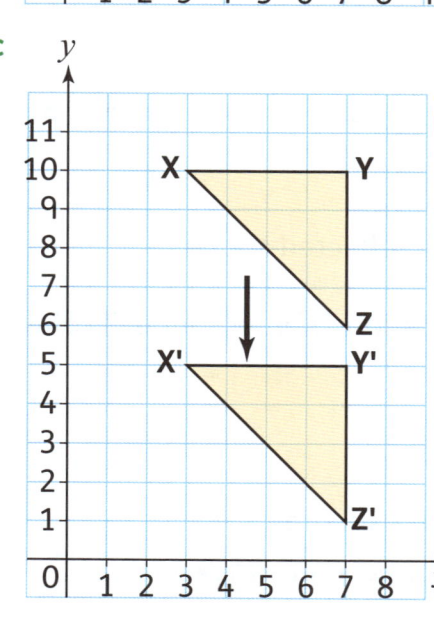

d

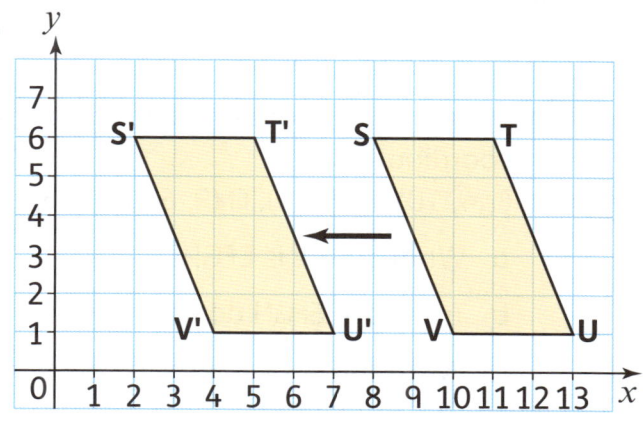

2 Write the coordinates of the vertices of each shape before and after translation.

➡ *Workbook page 68*

More moving shapes

1 Write the coordinates of the vertices of each shape after these translations.

   a 2 squares right

   b 4 squares down

   c 1 square up and 2 squares left

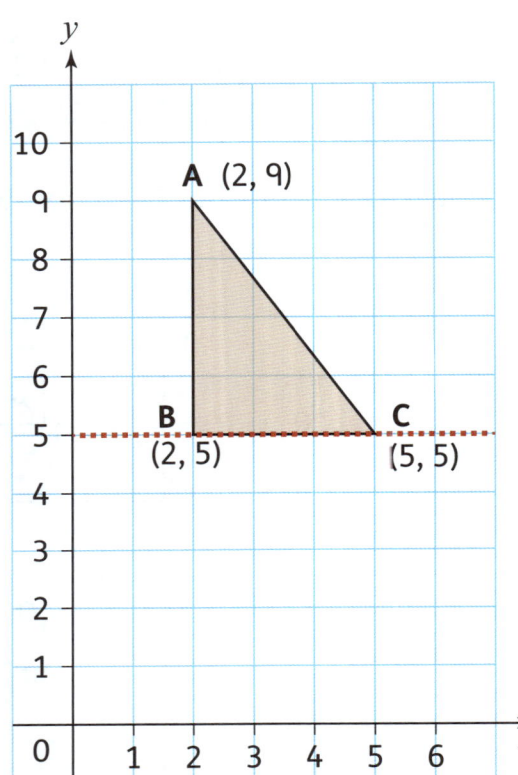 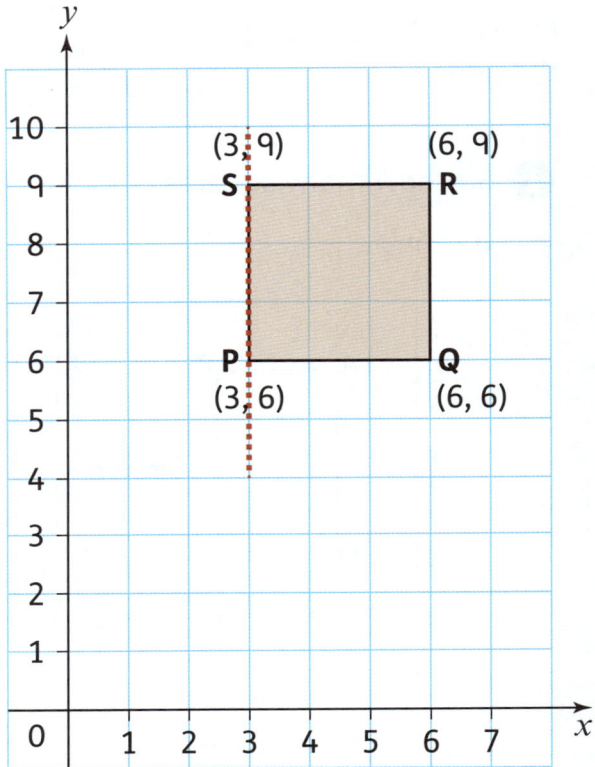

2 Reflect each shape in the red dotted mirror line.
Write the coordinates of the vertices of the reflected shape.

3 Draw a coordinate grid like the ones in question 1. Draw shape EFGH with coordinates:
E (4, 10), F (9, 10), G (9, 6), H (4, 6)

   a What shape is EFGH?

   b Translate EFGH down 3 squares and left 2 squares.

      Write the coordinates of the vertices of the shape after this translation.

   c Reflect shape EFGH using line GH as the mirror line.

Reflections and symmetrical patterns

We can reflect shapes in more than one mirror line to make **symmetrical** patterns. The mirror lines are lines of symmetry. If you fold the pattern along a mirror line, the two halves of the pattern fit exactly onto each other.

Reflect the single shape to make a pattern that is symmetrical about the vertical and horizontal mirror lines.

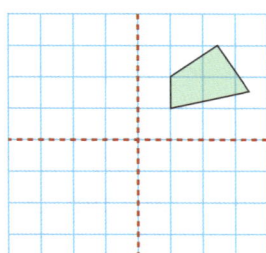

Start with this shape. The dotted lines are the mirror lines.

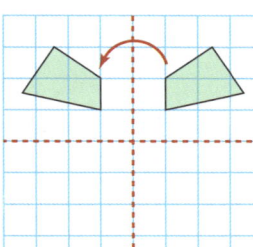

Reflect the shape in the vertical mirror line.

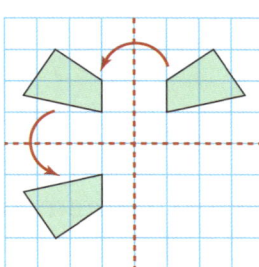

Reflect the second shape in the horizontal mirror line.

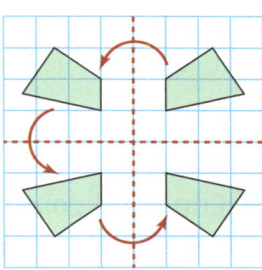

Reflect the third shape in the vertical mirror line.

Describe another way that you could reflect the shape to make this pattern.

1. Work in pairs to find the mistake in each of these reflection patterns.

a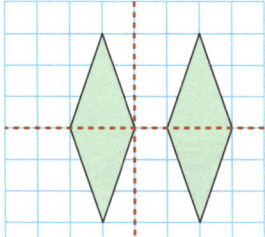

b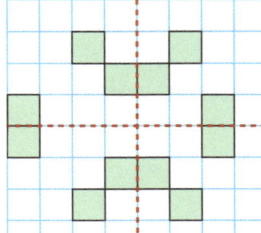

c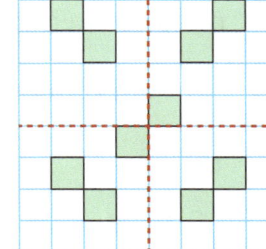

Problem solving

2. Pavel designed this square patterned tile.

   a How many mirror lines can you find in the design?

   b Design your own square tile with at least two mirror lines.

▶ *Workbook page 69*

Mixed practice 2

1 Which square number is closest to:

    a 60   **b** 70   **c** 130?

2 Which calculation is equivalent to 9^3?

    a 9×3 **b** $9 + 9 + 9$ **c** $9 \times 9 \times 9$

3 A pupil writes these two statements.

    $1^3 = 1$ $3^3 = 9$

    What mistake have they made? Can you find two ways to correct it?

4 Copy and complete these statements. Each set must contain a decimal, a percentage and two fractions.

    a $\frac{3}{4} = \square = \square = \square$ **b** $20\% = \square = \square = \square$

    c $0.25 = \square = \square = \square$ **d** $\frac{4}{10} = \square = \square = \square$

5 Calculate.

    a $\frac{3}{4}$ of an hour in minutes   **b** 20% of £15.00   **c** 0.25 of a metre

6 The distances on these road signs are all in kilometres, but Zara's car measures distance in miles. She uses this graph to help her to convert distances.

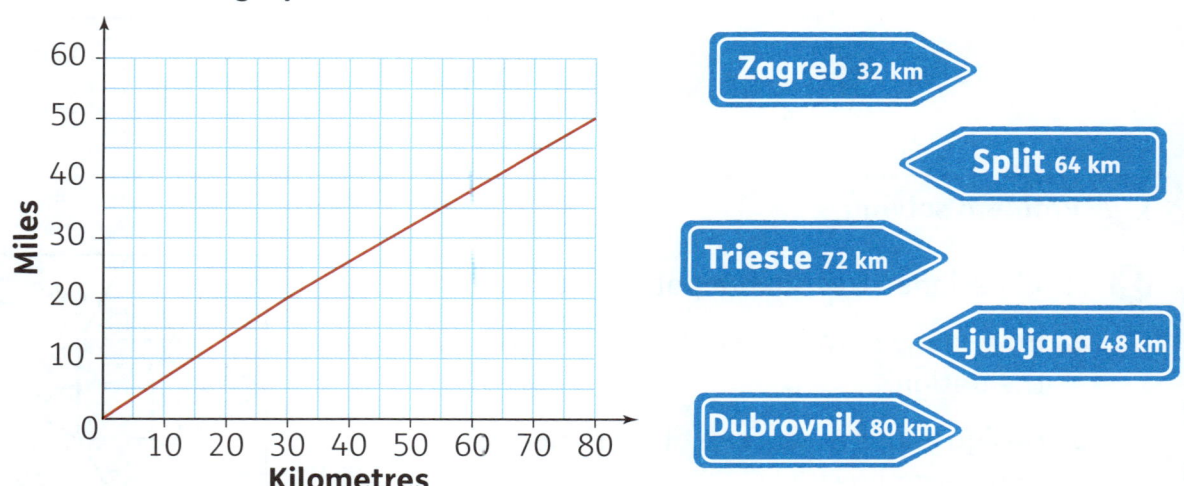

Conversion graph: kilometres–miles

Zagreb 32 km

Split 64 km

Trieste 72 km

Ljubljana 48 km

Dubrovnik 80 km

    a Look at the road signs next to the graph. Work out approximately how many miles it is to each place.

b It is 80 km to the next stop. Zara drives 40 miles. How many kilometres does she have left to drive?

c How can you use this graph to convert 100 km to miles?

7 Class A has 32 pupils and Class B has 28 pupils. $\frac{3}{4}$ of the pupils from each class go home by bus. How many pupils in these two classes do not go home by bus?

8 Solve each problem.

a I give $635 to each of my 4 children. How much do I give away altogether?

b Julia earns $375 each week. How much does she earn in 4 weeks?

9 Write three different additions, each with an answer of 1. Use fractions or decimals in all three calculations.

10 Write three different additions, each with an answer of 100. Use fractions in one sum, decimals in the next and whole numbers in the third.

11 Polygon ABCD is shown on a coordinate grid.

a What are the coordinates of point D?

b Which vertex is at (2, 5)?

c What type of polygon is this? How do you know?

d Why do points A and D have the same x-coordinates?

e Point E (4, 2) is plotted on the grid and joined to A and B with straight lines. What type of triangle is ABE?

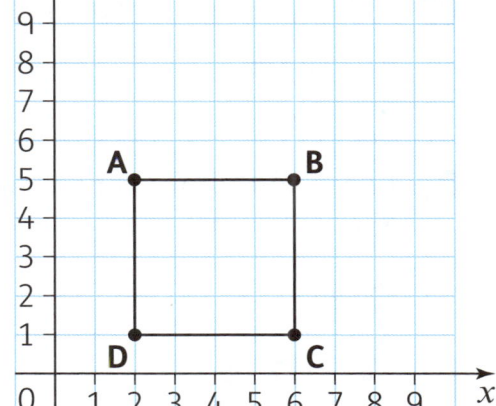

f Each square on the grid has sides 5 mm long. Calculate:

   i the perimeter of ABCD   **ii** the area of ABCD

g A pupil translates points A and B 1 square up and 2 squares left, to make a different polygon. What is the name of the new shape?

12 Marc has drawn a rectangle 40 mm long and 30 mm wide. He wants to increase its area by 12 cm². How could he do that? Draw a diagram as part of your answer.

UNIT 13 Multiplication and division 2

Written multiplication

> ### 💭 Think and share
>
> Compare these three methods of multiplying 427 × 3.
> Which method do you like best? Why?
> Why is an estimate useful?
>
> | H | T | O |
> |---|---|---|
> | ● ● ● ● | ● ● | ● ● ● ● ● ● ● |
> | ● ● ● ● | ● ● | ● ● ● ● ● ● ● |
> | ● ● ● ● | ● ● | ● ● ● ● ● ● ● |
>
> 1200   60   21 ⟶ 1281
>
> Estimate: 400 × 3 = 1200
>
> $$\begin{array}{r} 427 \\ \times \quad 3 \\ \hline 21 \\ 60 \\ 1200 \\ \hline 1281 \end{array}$$
> 3 × 7
> 3 × 20
> 3 × 400
>
> $$\begin{array}{r} 427 \\ \times \quad 3 \\ \hline 1281 \\ {\scriptstyle 2} \end{array}$$
>
> How else could you multiply 427 × 3? Share your ideas with your group.

1. Multiply using a written method. Estimate by rounding first.

   a 312 × 6   **b** 458 × 3   **c** 987 × 4

   d 1234 × 5   **e** 3123 × 7   **f** 4098 × 6

2. Two children have done their homework in different ways. Who has the correct answer for each calculation? What mistakes has the other child made?

   Freya

   a $\begin{array}{r} 3215 \\ \times \quad 3 \\ \hline 9645 \end{array}$ **b** $\begin{array}{r} 1441 \\ \times \quad 2 \\ \hline 4882 \end{array}$

   c $\begin{array}{r} 3315 \\ \times \quad 3 \\ \hline 9945 \end{array}$ **d** $\begin{array}{r} 3512 \\ \times \quad 6 \\ \hline 21072 \end{array}$

   Mina

   a $\begin{array}{r} 3215 \\ \times \quad 3 \\ \hline 9635 \end{array}$ **b** $\begin{array}{r} 1441 \\ \times \quad 2 \\ \hline 2882 \end{array}$

   c $\begin{array}{r} 3315 \\ \times \quad 3 \\ \hline 9936 \end{array}$ **d** $\begin{array}{r} 3512 \\ \times \quad 6 \\ \hline 21172 \end{array}$

➡ *Workbook page 70*

Multiply by 2-digit numbers

You can use the methods you already know to multiply any numbers.

32×16 Estimate: $30 \times 16 = 30 \times 10 + 30 \times 6 = 300 + 180 = 480$

Use a grid method

×	10	6	
30	300	180	480
2	20	12	32
			512

Use a written column method

```
   32
 × 16
   12     6 × 2
  180     6 × 30
   20     10 × 2
  300     10 × 30
  512
    1
```

```
   32
 × 16
  192
   1
  320
  512
    1
```

---

1 Use a written method to multiply. Estimate the product first.

a 15×13 b 20×45 c 25×37

d 18×66 e 19×39 f 23×86

g 66×12 h 54×23 i 19×91

2 Work out how many maize plants there are on each farm.

a    b    c    d

27 rows   92 rows   61 rows   81 rows
55 plants per row   23 plants per row   50 plants per row   49 plants per row

---

Problem solving

3 Nelson is a delivery driver. He delivers
525 parcels per week on average.
For 3 weeks each year he is on holiday.
How many parcels does he deliver in a year?

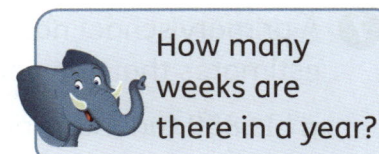

How many weeks are there in a year?

➡ *Workbook page 71*

Practise multiplying

1 Calculate.

a 299×10	**b** 25×19	**c** 19×35
d 125×4	**e** 624×2	**f** 127×9
g 299×8	**h** 350×6	**i** 12×21

2 Work out how many or how much.

a Nisha has 427 stickers. Pete has 4 times as many.

b Farmer Joe has 327 cows. Farmer Mavis has twice as many.

c Palesa planted 6 rows with 126 plants in each.

d Pete has $419. His brother has 5 times as much.

e Samantha has 269 stickers. Sandra has 10 times as many.

f Josh saved $124, but he needs 8 times as much to buy a car.

g A school ordered 12 sets of 65 blocks.

h A stadium has 65 rows of 42 seats.

3 378 cars drive over a bridge each hour. How many cars drive over the bridge:

a in 8 hours

b in 12 hours?

4 Jessie's school is 381 metres from her house. She walks to school and back each day.

a How far does she walk to school and back in a day?

b How far does she walk in total in 5 days?

5 Tayo charges $83 for a trip to the airport. How much does he earn for 55 trips?

6 Robert says 526 times 3 is 1568. Is he correct?

7 A primary school has 27 classes with 32 pupils in each. The principal estimates that she needs to order 850 chairs for the pupils.

a Is 850 too many or too few chairs?

b How should the principal change her order?

8 A leaking water tank loses 89 litres per day. How much water will it lose in:

a 10 days

b 2 weeks?

Revisit division

You can use multiplication and division facts to help you divide mentally.

These are some of the things that you have already learnt about division.

You can think of division as repeated subtraction.
To work out $18 \div 3$, you can ask
yourself: 'How many 3s make 18?'

How can this help you to work out $180 \div 3$?

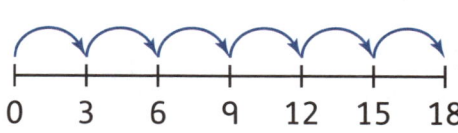

Division is the **inverse** of multiplication.

To work out $42 \div 6$, you can ask yourself: 'What multiplied by 6 will give me 42?'

$\boxed{} \times 6 = 42$ You know from your times table facts that $6 \times 7 = 42$.

So, $42 \div 6 = 7$
What is $420 \div 7$? Why?

You can use place value to divide by 10 and 100.

When you divide by 10, the digits move 1 place to the right.
$110 \div 10 = 11$
When you divide by 100, the digits move 2 places to the right.
$4200 \div 100 = 42$

1 Do these divisions.

    a $16 \div 4$ **b** $16 \div 8$ **c** $16 \div 2$ **d** $48 \div 6$ **e** $48 \div 8$ **f** $48 \div 2$

    g $64 \div 8$ **h** $64 \div 4$ **i** $64 \div 2$ **j** $45 \div 5$ **k** $45 \div 9$ **l** $45 \div 3$

    m $90 \div 10$ **n** $90 \div 5$ **o** $90 \div 2$ **p** $100 \div 10$ **q** $100 \div 100$ **r** $100 \div 50$

2 There are 48 children in a class. The teacher wants to arrange them into 8 equal groups. How many children will there be in each group?

3 Callie has 54 sweets to share equally among 6 friends. How many sweets will each friend get?

Workbook page 72

Written division

You can use written methods to divide larger numbers. One method is called **short division**.

You write short divisions with a 'division house' and divide the numbers one by one.

$98 \div 7 \longrightarrow$ $7\overline{)9^28}$ with 1 above $9 \div 7 = 1 \text{ r } 2$ $\longrightarrow$ $7\overline{)9^28}$ with 14 above $28 \div 7 = 4$

$544 \div 4 \longrightarrow$ $4\overline{)5^144}$ with 1 above $5 \div 4 = 1 \text{ r } 1$ $\longrightarrow$ $4\overline{)5^14^24}$ with 13 above $14 \div 4 = 3 \text{ r } 2$ $\longrightarrow$ $4\overline{)5^14^24}$ with 136 above $24 \div 4 = 6$

If the first number is less than the number you are dividing by, divide into the first two numbers.

$159 \div 3 \longrightarrow$ $3\overline{)159}$ with 5 above $15 \div 3 = 5$ $\longrightarrow$ $3\overline{)159}$ with 53 above $9 \div 3 = 3$

Write the answer above the 5.

1 Estimate first. Then divide and show your working.

a $85 \div 5$	b $96 \div 6$	c $76 \div 4$
d $92 \div 4$	e $84 \div 6$	f $84 \div 7$
g $96 \div 8$	h $57 \div 3$	i $91 \div 7$

2 Estimate first. Then divide and show your working.

a $156 \div 6$	b $282 \div 3$	c $564 \div 4$
d $144 \div 6$	e $155 \div 5$	f $176 \div 8$
g $196 \div 4$	h $208 \div 8$	i $195 \div 5$

3 Mrs Singh wants to put 132 eggs into boxes.
Each egg box holds 6 eggs. How many boxes can she fill?

Estimate and divide with remainders

You already know that some divisions leave a **remainder**.
We write the remainder as part of the answer.

How many times does 4 go into 169?
Estimate: 4 × 40 = 160, so the answer should be close to 40.

$$\begin{array}{r} 42 \text{ r } 1 \\ 4\overline{)169^1} \end{array}$$

169 ÷ 4 = 42 r 1

1 Estimate first. Then divide. Use the method you find easiest.

a 407 ÷ 3	**b** 572 ÷ 4	**c** 577 ÷ 4
d 325 ÷ 4	**e** 253 ÷ 5	**f** 217 ÷ 3
g 169 ÷ 2	**h** 400 ÷ 9	**i** 256 ÷ 6
j 500 ÷ 7	**k** 328 ÷ 7	**l** 649 ÷ 8

2 A coach needs 7 pupils to make each netball team.

a There are 136 pupils. How many teams can the coach make?

b How many pupils will be left over?

3 Sammy has 199 metres of rope.
He wants to cut it into 4 m lengths.

a How many 4 m pieces can he cut?

b How much rope will be left over?
Write this in centimetres.

4 A farmer has 237 maize plants. He wants to plant them in 3 rows. How many plants should he put in each row?

Can you divide the remainder?

What you do with a remainder depends on the situation.
Sometimes you can write the remainder as a fraction or a decimal. At other times it makes sense to leave it as a whole number.

These 9 children were divided into 2 teams. One person was left over. You cannot divide up a person!

$9 \div 2 = 4 \text{ r } 1$

Sometimes you can divide the remainder into smaller pieces; sometimes you cannot.

9 cakes are shared between 2 children. The last cake can be shared between the 2 children by cutting it in half.

$9 \div 2 = 4\frac{1}{2}$ cakes

1. Do a division for each situation. Decide when to leave a remainder and when to write the remainder as a fraction.

    a Share 12 marbles between 5 children.

    b Divide 3 lettuces among 2 rabbit hutches.

    c Share 15 cakes between 4 plates.

    d Give 20 children an equal share of 25 metres of ribbon.

    e Divide 40 counters between 13 children.

    f Share 500 grams of cake mix between 8 baking tins.

> If it helps, draw a diagram.

Problem solving

2. 145 pupils attended a sporting event. For safety reasons, they were only allowed into the event in groups of 8.

    a How many groups should the pupils be divided into?

    b What do you think should happen to the remaining pupils?

3. Charlotte sells fruit in bags. She can buy bags that hold 6, 8 or 10 pieces of fruit. She wants to pack 438 pieces of fruit into bags with the fewest possible pieces of fruit left over. What size bags should she buy?

Workbook page 73

More work with remainders

💡 Problem solving

Draw a bar model to help you solve each problem.

1. 79 people are divided into 4 equal groups.

   a. How many people are in each group?

   b. How many people are left over?

2. Some pupils are making a number line on the ground in the school yard.

   The number line is 10 m long.
   The pupils divide it into 8 equal parts.
   How long is each part?

3. Malik has 3 litres of juice. He wants to pour equal amounts into 4 jugs. How much juice should he pour into each jug? Give your answer as a fraction or decimal.

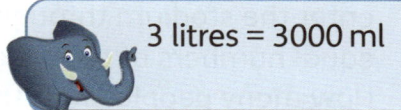

3 litres = 3000 ml

4. Share £77 pounds between 4 people. How much will each person get?

5. Nadja has 1436 sweets to pack into bags. Each bag must have a minimum of 3 sweets and no more than 8 sweets. Each bag must have the same number of sweets. She wants to have the fewest possible loose sweets left over. How can she pack the sweets?

6. A building has 2 elevators. Each elevator can hold a maximum of 9 people. On one day, 380 people use the elevators. What is the fewest number of times that each elevator is used that day?

7. Read what two pupils said about a division calculation.

   My answer is between 70 and 75, and the remainder is less than 4.

   My answer is between 90 and 100, and the remainder is more than 5 but less than 9.

   Write two possible divisions that each pupil could have done.

More division

1 Estimate and then calculate the answers. Show your method.

a 804 ÷ 4   **b** 309 ÷ 3   **c** 500 ÷ 5

d 999 ÷ 9   **e** 639 ÷ 3   **f** 602 ÷ 2

g 770 ÷ 7   **h** 282 ÷ 2   **i** 936 ÷ 3

j 600 ÷ 5   **k** 990 ÷ 3   **l** 900 ÷ 9

 Problem solving

2 I have $906. I share it equally between 3 envelopes.
How much do I put into each envelope?

> If there is a remainder, think about what you should do with it.

3 840 people attend a sports match. They enter the stadium through 4 gates, with equal numbers of people entering through each gate.
How many people enter through each gate?

4 I have 2 boxes of stickers. Each box contains the same number of stickers. There are 624 stickers in total.

  a How many stickers are there in each box?

  b There are 8 sheets of stickers in each box. How many stickers are there on a sheet?

5 A teacher shares 147 sheets of paper equally between 5 groups.
How many sheets of paper does each group get?
How many are left over?

6 In a pottery class, 300 kg of clay is shared equally between 9 pupils.
How much clay does each pupil get?

7 On a hike, 6 people have to carry 15 litres of water between them.
They each carry the same amount.
How much water does each person carry?

Solve division problems

1 500 metres of cable are cut into 6-metre lengths. How many lengths are there?

Choose the best method to solve each problem. Use mental strategies where you can.

2 126 cm of fabric was left at the end of a roll. The shop owner cut it into 5 equal scraps. How long was each scrap of fabric?

3 How many 8-page sections are put together to make a book of 448 pages?

4 Books are packed into boxes of 10. How many boxes are needed for 264 books?

5 Jo needs $450 to pay for a holiday.
She saves $8 a week.
How many weeks will it take her to save enough money?

6 Sally wants to save her photographs onto her hard drive. Each photo is 4 megabytes in size. She has 50 megabytes of memory free on her hard drive. How many photographs can she save?

7 Carly needs to print her school newsletters.
Each newsletter is 4 pages long. A ream of paper holds 500 pages.
How many newsletters can she print (single-sided) from 1 ream?

8 Nika did these divisions on a whiteboard. Some digits got smudged.
Work out what the smudged digits could be in each calculation.

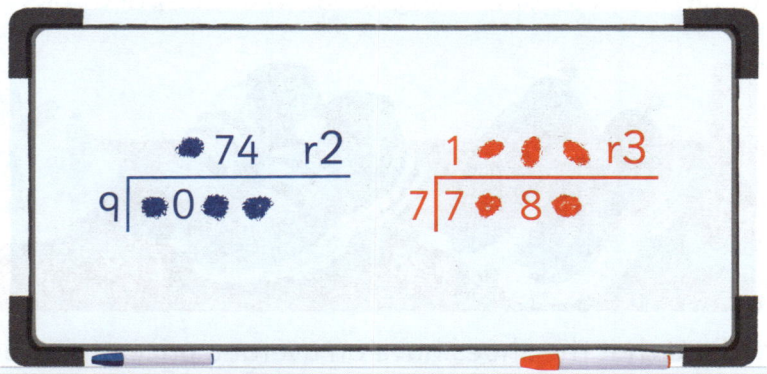

Mixed problems

1 For each estimated answer, choose the most suitable calculation.
Tell your partner how you decided.

Estimate	Calculations		
a 1100	$1000 \div 9$	100×9	$10\,000 \div 9$
b 400	40×11	420×11	4×120
c 100	$730 \div 7$	$7000 \div 20$	77×10
d 4000	412×100	$100\,000 \div 4$	$44\,000 \div 10$

2 Jamila bought 24 packs of beads. Each pack cost £4 and contained 72 beads. She used the beads to make 8 necklaces, each with the same number of beads.

  a How many beads did Jamila buy?

  b How many beads did she use for each necklace?

  c What do you think a fair price would be for each necklace? Give reasons for your answer.

3 A school ordered books costing £315, 10 boxes of paper costing £24 each and a whiteboard for £745. They paid for these items in 4 equal weekly amounts. How much did they pay each week?

4 Jay bought these 3 pairs of running shoes in a half-price sale.

-SALE-
ALL SHOES
½
MARKED PRICE

£38   £28   £45

Jay tells a friend that the shoes have an average price of £18.50 per pair. Explain how Jay worked this out.

What is a rate?

A **rate** compares two different quantities.
Speed is a rate that compares distance travelled with time.

If a car is travelling at 60 kilometres per hour, this means that it travels a distance of 60 kilometres in 1 hour. We write 60 km/h.

To solve problems involving rates, you divide or multiply both quantities by the same number.

Look at these examples.

A car travels 280 km in 4 hours.
Work out its speed in km/h.
To get the speed in 1 hour, we divide both quantities by 4.
4 hours ÷ 4 = 1 hour
280 km ÷ 4 = 70 km
The speed is 70 km/h.

Li's fitness tracker records her heart rate in beats per minute.
Her heart rate is 68 beats/minute. How often will her heart beat in 5 minutes?
1 minute × 5 = 5 minutes
68 beats × 5 = 340 beats
Her heart will beat 340 times in 5 minutes.

1. Work with a partner. Discuss these units for rates. Try to think of two examples where you would use each rate.

pounds per gram   words per minute   kilometres per litre

litres per minute   metres per second   dollars per hour

pence per metre   pounds per day

2. A car travels 560 km in 5 hours.

   a What is its speed in km/h?

   b How far will it travel in 8 hours at this speed?

3. A library charges £1.15/week for overdue books. What does it charge for a book that is 21 days overdue?

4. 8 markers cost £3.20. How much does 1 marker cost?

5. Train A travelled at 90 km/h for 5 hours. Train B travelled at 110 km/h for $4\frac{1}{2}$ hours. Which train travelled the furthest?

Work with negative numbers

Positive and negative numbers

☁️ Think and share

- What does 'Level 0' represent on the drawing?
- How many floors are above ground level?
- How many are below ground level?
- How could you describe which floor the green car is on, using numbers?
- How could you describe the top and bottom floors using numbers?

Work in pairs to draw a set of buttons for the elevators in this building, using only numbers and symbols.

Numbers greater than 0 are **positive numbers**. 1, 2, 3 and 4 are positive numbers.
Numbers less than 0 are **negative numbers**. −1, −2, −3 and −4 are negative numbers.
We always write a minus sign (−) when the number is negative.
Positive numbers don't need a sign, but they can be written as +1, +2 and so on.
0 is neither negative nor positive.

LEVEL 0

1 Look at the picture.

a What is the approximate depth of the two fish swimming together?

b A bird is flying 3 m above sea level. Approximately how far would the bird need to dive in order to catch the fish closest to the surface?

sea level

0 m

−1 m

−2 m

−3 m

Count on and back through 0

If you count along a standard number line from left to right, the numbers increase in value.

If you count back from right to left, the numbers decrease in value.
For example, −3 < 1 (negative 3 is less than 1).

If you start at 3 on this number line and count back 5, you land on −2.

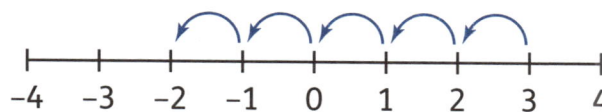

-4 -3 -2 -1 0 1 2 3 4

1 Use this number line. Follow the directions. Write the number you land on.

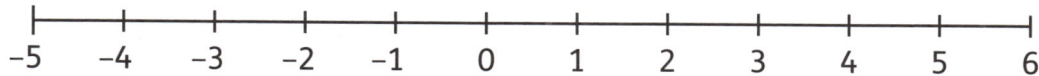

-5   -4   -3   -2   -1   0   1   2   3   4   5   6

a Start on 3, count back 6.
b Start on −4, count on 3.

c Start on −2, count on 7.
d Start on 5, count back 5.

e Start on −1, count on 2.
f Start on 2, count back 4.

g Start on 4, count back 2.
h Start on −5, count on 4.

2 Copy the statements. Fill in < or > to compare the numbers.

a −4 ☐ −2
b 3 ☐ −1

c −1 ☐ −3
d 2 ☐ 5

e −4 ☐ 2
f −2 ☐ 0

3 Look again at this picture of a building.

a Make up five questions about it using counting on or back through 0.

b Share your questions with a partner and answer each other's questions.

LEVEL 0

➡ *Workbook page 74*

Count in steps

You already know how to count forwards and backwards in steps.

Count forwards in 10s:	10	20	30	40	50
Count back in 300s:	4500	4200	3900	3600	3300

You can count backwards and forwards through 0 using negative numbers.

Count back in 2s:	6	4	2	0	−2	−4
Count back in 10s:	30	20	10	0	−10	−20

1 Start at 20. Count back in the given steps until you are one step past 0. List the numbers you count.

    a Count in sixes.   **b** Count in tens.   **c** Count in fours.

    d Count in twelves.   **e** Count in twenty-fives.   **f** Count in fifteens.

2 Work out the term-to-term rule for each sequence. Copy each sequence and fill in the missing numbers.

    a 135, 124, ☐, 102, ☐, ☐   **b** 19, 13, ☐, ☐, ☐, −11

    c −4, −2, 0, ☐, ☐, 6   **d** −25, −15, ☐, 5, 15, ☐

3 Write the first ten numbers in the sequence for each of these rules.

    a Start at 55. Count back in tens.   **b** Start at 70. Count back in twenties.

    c Start at −40. Count forwards in fives.   **d** Start at −500. Count forwards in hundreds.

Problem solving

Sea level is at 0 metres.

4 A diver stops every metre to equalise pressure in her ears as she dives down to a reef.

    a She stops 13 times on the way down to the reef. At what depth is the reef?

    b On the way back up to the boat, she stops for 5 minutes, 8 m above the reef. At what depth does she stop?

➡ *Workbook page 75*

Number sequences

The Caspian Sea is an inland sea in Asia.
The surface of the sea is 28 m below sea level.

A scientist sends a camera down from the surface. It is programmed to stop every 2 m as it descends.

We can write this as a number sequence:
−28, −30, −32, −34, −36, ...
The rule for this sequence is: start at −28 and count back in twos.

After the camera reaches a depth of −40 m, the scientist programmes it to stop every 4 m as it rises, to take photos.

> A number sequence is a set of numbers that follow a rule.

We can write these depths as a number sequence like this: −40, −36, −32, −28.

The rule for this sequence is: start at −40 and count on in fours.

Why does this sequence have to stop at −28?

1 Work out the term-to-term rule and the next three numbers in each sequence.

a −10, −8, −6, −4, ...

b −20, −15, −10, −5, ...

c −40, −20, 0, 20, ...

d 5, 0, −5, −10, ...

e −125, −150, −175, −200, ...

f 500, 400, 300, 200, ...

g 120, 50, −20, −90, ...

h 180, 100, 20, −60, ...

2 The table gives the depth below sea level of some of the lowest places on Earth.

Place	Dead Sea	Lake Assal	Qattara Depression	Denakil Depression	Laguna del Carbón
Location	Jordan	Djibouti	Egypt	Ethiopia	Argentina
Depth	−414 m	−155 m	−133 m	−125 m	−105 m

a Work in pairs. Make up a counting sequence using each depth. Write the first five numbers in your sequence.

b Swap sequences with another pair. Work out the rules they used.

Temperature changes

This thermometer shows the temperature in degrees Celsius. The short way of writing degrees Celsius is °C. We show temperatures lower than 0 °C as negative numbers.

> On the thermometer scale, one division represents 2 degrees. This thermometer shows a temperature of 4 degrees below 0. We read this as minus 4 degrees Celsius. We write −4 °C.

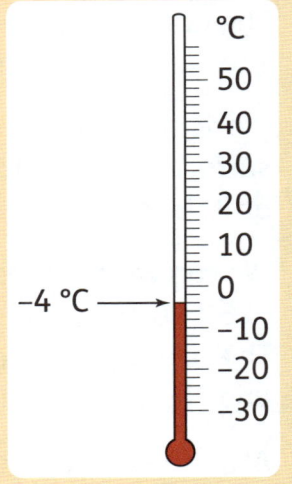

You can use the scale on the thermometer like a number line to work out changes in temperature.

1 Here are five thermometers. For each one:

    a Write the temperature that is 10 °C warmer.

    b Write the temperature that is 12 °C colder.

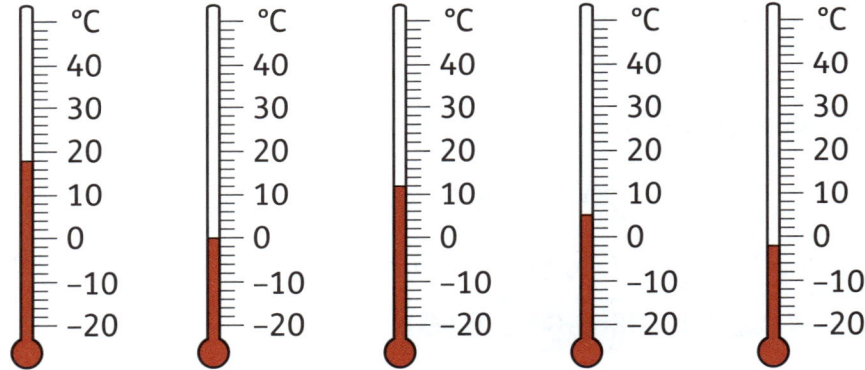

2 One night in Calgary, Canada, the temperature dropped to 8 degrees below 0.

    a What temperature did the thermometer show?

    b By noon, the temperature had risen 15 degrees. What was the new temperature?

    c By 8 p.m. that night, the temperature had dropped 7 degrees since noon. What was the temperature then?

    d At midnight, the temperature was −3 °C. How much had it dropped since 8 p.m.?

➡ *Workbook page 76*

Add and subtract with negative numbers

A number line can help you to add and subtract with negative numbers.

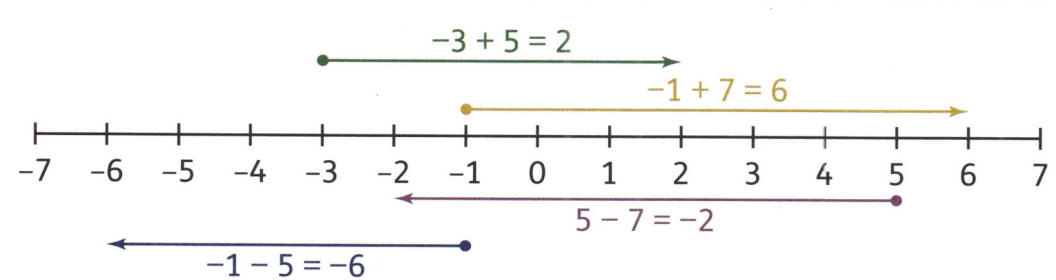

Misha says that you cannot subtract a greater number from a smaller one. Explain why this is incorrect. Use the diagram to show what you mean.

What does it mean if someone writes 4 + (−5)? What would the answer be? Why?

1 Calculate. Use the number line above to help you.

    a 2 − 3   b −3 + 1   c −4 + 3   d −1 − 4

    e −6 − 6   f 6 − 7   g −6 + 6   h −6 + (−1)

2 Copy the number sentences. Fill in the missing values.

    a −4 + ☐ = 10   b 2 + ☐ = −13   c ☐ − 9 = −11

    d −13 + ☐ = −4   e 7 − ☐ = −14   f 9 + ☐ = 0

 Problem solving

 There are 1000 m in 1 km.

3 The lowest point on Earth is the bottom of the Mariana Trench in the Pacific Ocean. The highest point is the top of Mount Everest in the Himalayas. Look at the diagram.

    a If you travelled from the top of Mount Everest to the bottom of the Mariana Trench, how many metres would you travel?

    b A small submarine travels down into the trench to a depth equal to the height of Mount Everest. How far is that from the bottom of the trench?

8.9 km ----- ⌐

Mount Everest

sea level

Mariana Trench

----- −10.9 km

Calculate with decimals

Pairs that make 1

Think and share

These are the addition facts for whole numbers to 10.

$1 + 9 = 10$ $2 + 8 = 10$ $3 + 7 = 10$ $4 + 6 = 10$ $5 + 5 = 10$

The strip represents 1 whole. Each part of the strip is $\frac{1}{10}$ or 0.1 of the whole.

$\frac{3}{10}$ or 0.3 of the strip is shaded.

Each of these shapes represents 1 whole.

What fraction of each shape is shaded?
Give your answers as decimals.
Say one decimal addition and one
subtraction fact for each diagram.

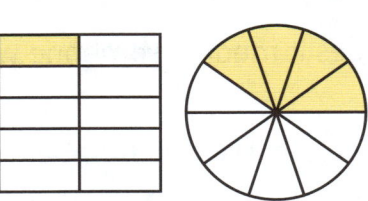

Explain how you can use number facts
to add tenths to make 1 whole.
How can you use the facts to make a related set of subtraction facts?

1 Copy and complete these number sentences.

　　a $0.1 + \square = 1$　　b $1 - 0.7 = \square$　　c $1 - 0.5 = \square$　　d $\square + 0.8 = 1$

　　e $1 - 0.6 = \square$　　f $1 - 0.2 = \square$　　g $0.3 + \square = 1$　　h $1 - \square = 0.8$

　　i $\square + 0.6 = 1$　　j $0.5 + 0.5 = \square$　　k $1 - \square = 0.1$　　l $1 - 0.5 = \square$

Problem solving

What fraction of a
bar is each part?

2 Naadira needs 5 chocolate bars this
size to bake chocolate brownies.

She has these broken pieces.
Does she have enough to make
the brownies?

Explain how you decided.

Use facts to add hundredths

You know that $34 + 66 = 100$.   You can use that to work out that $0.34 + 0.66 = 1.00$.

0.34 is 34 hundredths and 0.66 is 66 hundredths.
Together they make 100 hundredths, which is the same as 1 whole.

This square has been divided into 100 squares.

Each square is $\frac{1}{100}$ or 0.01 of the whole.

34 parts are shaded. This is 0.34 of the whole.
66 parts are not shaded. This is 0.66 of the whole.
$0.34 + 0.66 = 1$ whole.

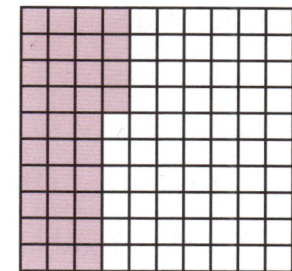

Addition and subtraction are inverse operations.
$1 - 0.34 = 0.66$ and $1 - 0.66 = 0.34$

1. Copy and complete each set of number sentences.

   a $27 + \boxed{} = 100$ $0.27 + \boxed{} = 1$ $1 - 0.27 = \boxed{}$

   b $55 + \boxed{} = 100$ $0.55 + \boxed{} = 1$ $1 - 0.55 = \boxed{}$

   c $83 + \boxed{} = 100$ $0.83 + \boxed{} = 1$ $1 - 0.83 = \boxed{}$

   d $39 + \boxed{} = 100$ $0.39 + \boxed{} = 1$ $1 - 0.39 = \boxed{}$

2. For each length of ribbon, say how much more is needed to make 1 metre.

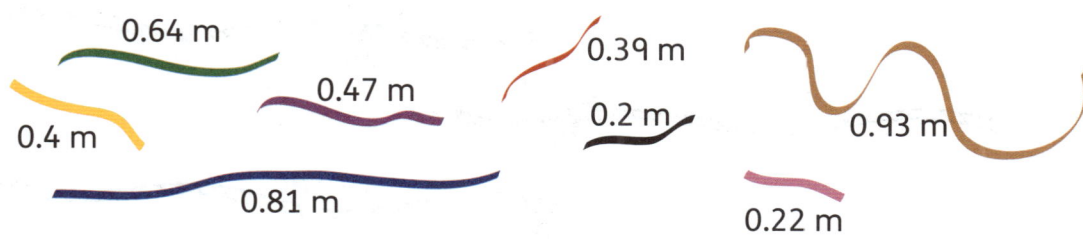

0.64 m

0.47 m

0.39 m

0.2 m

0.4 m

0.81 m

0.93 m

0.22 m

3. Look at this number line.

   a   Write the decimal shown by each arrow.

   b   Next to each decimal, write what you need to add to make 3.

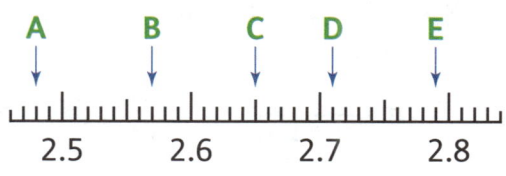

➡ *Workbook page 77*

Add and subtract across a whole number

How can you work out how long this chameleon is from the end of its nose to the end of its tail?

Look at these methods for adding the two lengths.

body (including head) = 7.7 cm long
tail = 9.5 cm long

7.7 is 77 tenths   9.5 is 95 tenths

77 + 95 = 172 tenths, which we write as 17.2

The chameleon is 17.2 cm long.

We can add 9.5 in parts on a number line.

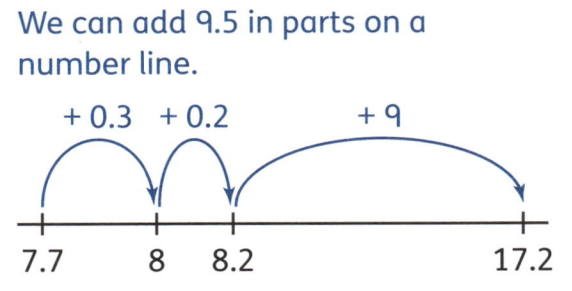

A different chameleon is 18.3 cm long. Its tail is 11.6 cm long. Discuss how you would work out how long its body is.

1. Calculate the total length of each piece of rope in metres.

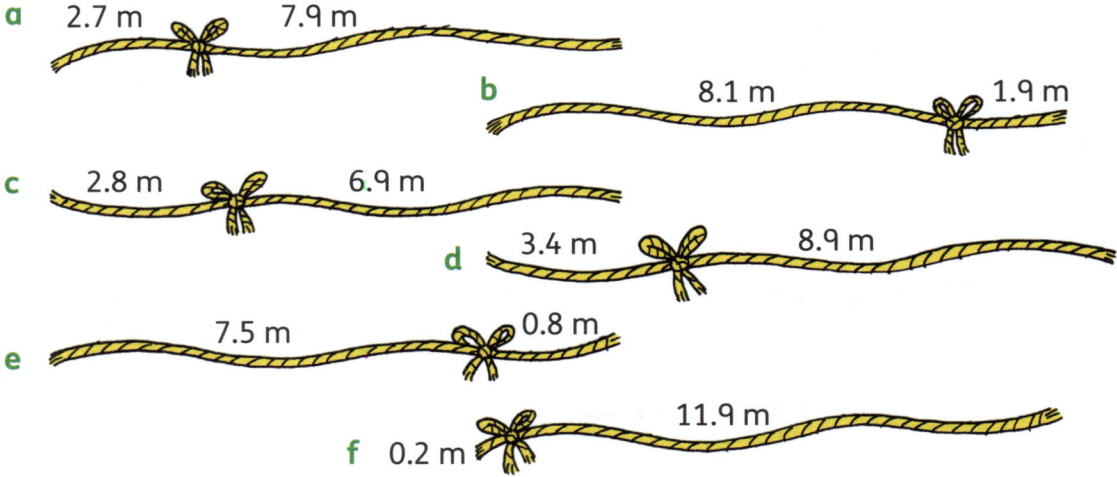

a   2.7 m   7.9 m

b   8.1 m   1.9 m

c   2.8 m   6.9 m

d   3.4 m   8.9 m

e   7.5 m   0.8 m

f   0.2 m   11.9 m

2. Work out the difference in mass for each pair of objects. Give your answers in kilograms.

a   2.3 kg   1.9 kg

b   4.7 kg   2.9 kg

c   4.1 kg   3.8 kg

Workbook page 78

Add and subtract any decimals

Look at these methods of adding and subtracting decimals.

$8.2 + 9.8$ $= 8 + 0.2 + 9 + 0.8$
$= (8 + 9) + (0.2 + 0.8)$
$= 17 + 1$
$= 18$

Place value and number facts

$8.2 - 0.7$

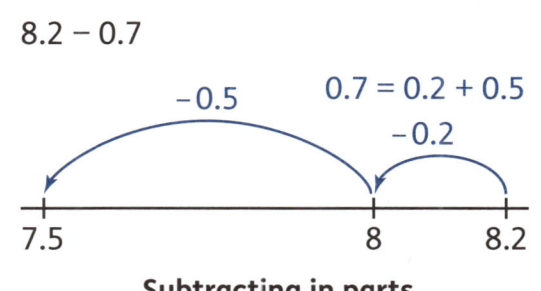

$0.7 = 0.2 + 0.5$

Subtracting in parts

$9.7 - 4.4$

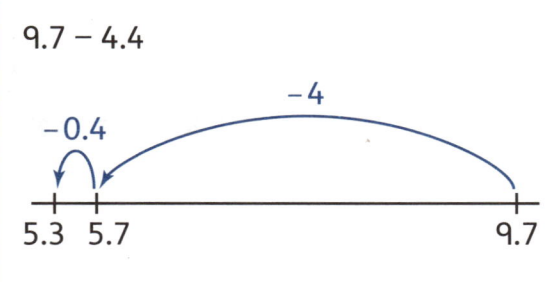

A number line

Or $9 - 4 = 5$ $0.7 - 0.4 = 0.3$ 5.3

$12.23 + 9.5$

$\begin{array}{r} 12.23 \\ + \ 9.50 \\ \hline 21.73 \\ \hline \small{1} \end{array}$

Make sure the decimal places are lined up correctly.

Write 0 as a place holder for empty tenths or hundredths.

Written methods

Which methods are easiest to understand? Why?
Which methods would you use? Why?

1. Add. Show your working.

   a $2.13 + 3.14$
   b $3.20 + 1.59$
   c $3.4 + 2.9$
   d $12.25 + 34.24$
   e $32.12 + 19.45$
   f $12.09 + 9.98$

2. Subtract. Show your working.

   a $6.96 - 3.45$
   b $9.77 - 3.25$
   c $8.95 - 4.64$
   d $5.43 - 2.39$
   e $9.53 - 6.17$
   f $8.14 - 6.51$

Problem solving

Write money amounts with 2 decimal places.

3. A pair of shorts costs £8.45 and a T-shirt costs £4.99.

   a What does it cost to buy a pair of shorts and a T-shirt?
   b What is the difference in price between the two items?
   c How much change would you get from £15 if you bought both items?

➡ *Workbook page 79*

Multiply and divide by 10 and 100

In Unit 7 you learnt how to multiply and divide whole numbers by 10 and 100 using place value and mental strategies.

You can use the same strategies to work with decimals.

Look at these examples.

Discuss the patterns that you can see.

How can these help you to multiply and divide decimals by 10 and 100?

$15 \times 10 = 150$ $120 \div 10 = 12$

$1.5 \times 10 = 15$ $1.2 \div 10 = 0.12$

$15 \times 100 = 1500$ $120 \div 100 = 1.2$

$1.5 \times 100 = 150$ $1.2 \div 100 = 0.012$

1. Work with a partner. Look at the number chains. Work out which operation gets you from one number to the next in each chain. For example, to get from 0.5 to 5 you would multiply by 10.

a

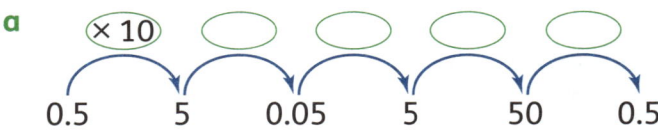

0.5   5   0.05   5   50   0.5

b

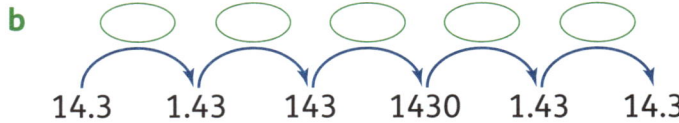

14.3   1.43   143   1430   1.43   14.3

c

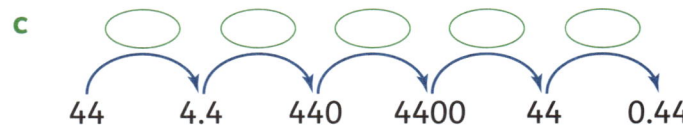

44   4.4   440   4400   44   0.44

2. Make two number chains of your own. Swap with a partner and complete each other's number chains. You can leave out the × or ÷ signs *or* the numbers.

3. Do these calculations mentally. Write the answers only. Use a calculator to check your work.

    a 3.4×100 **b** 5.7×100 **c** 0.8×10

    d $0.2 \div 10$ **e** $12.7 \div 10$ **f** $410 \div 100$

    g 0.007×100 **h** 0.04×10 **i** $2.3 \div 100$

Multiply decimals by whole numbers

$31 \times 3 = 93$ What is 3.1×3?

Estimate first:
3.1 rounds to 3
$3 \times 3 = 9$
So, 3.1×3 must be close to 9.

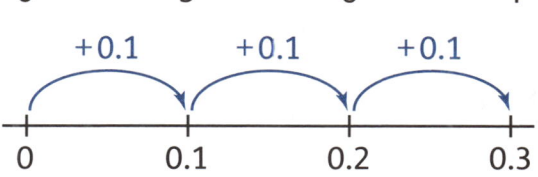

Work it out like this:
$3.0 \times 3 = 9.0$
$0.1 \times 3 = \underline{0.3}$
$\underline{9.3}$

When you write your calculations like this, remember to keep the decimal points lined up underneath each other.

1. Estimate the answers to these first. Then calculate.
   Compare your estimate with the answer.

   a 1.8×3 **b** 5.1×5 **c** 2.3×4 **d** 6.9×2

   e 9.9×6 **f** 4.5×3 **g** 7.7×4 **h** 2.2×8

 Problem solving

Remember to show your working.

2. Estimate each answer and then solve the problem.

   a Mrs Downes collected $2.50 from each of 8 children. How much did she collect altogether?

   b 6 children each made a paper chain decoration 1.5 m long. When the pieces were joined together, how long was the whole decoration?

   c A car travelled at a speed of 91.4 km/h. How far did it travel in 4 hours?

   d Carpet costs £25.99 per metre. How much does it cost for 10 metres?

➡ *Workbook page 80*

More multiplying

1 Estimate then calculate.

   a 3×4.2 **b** 5×6.1 **c** 9×3.1

   d 5.2×4 **e** 6.9×7 **f** 4.4×2

   g 2.4×9 **h** 9.3×7 **i** 8.4×5

Problem solving

2 Here are four rows of tiles. The length and width of one tile in each row is given. The tiles are not drawn to scale, but all the tiles in each row are the same size.

22.4 cm

11 cm

4.5 cm

2 cm

45.2 cm

25 cm

55.5 cm

15 cm

   a Work out the length of each row of tiles.

   b What area is covered by each row of tiles?

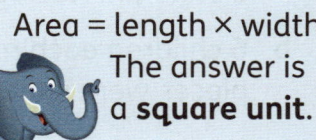

Area = length × width
The answer is
a **square unit**.

3 This is Kate's recipe for one bowl of fruit salad.

Work out how much of each fruit she needs to make:

   a 2 bowls

   b 5 bowls

   c 8 bowls

1.2 kg watermelon

1.7 kg apple

1.1 kg pineapple

0.5 kg grapes

1.8 kg mango

➡ *Workbook page 81*

Mixed decimal problems

Work out what operation you need to do. Then estimate the answer before you calculate.

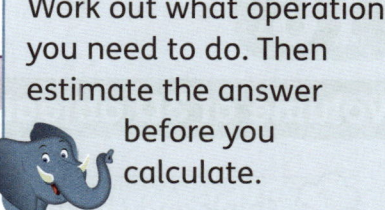

Problem solving

1 Mike buys 7 planks that are 1.3 m long.

   He pays $9 per metre for the planks.

   a   What is the total length of all the planks?

   b   What does it cost for all the planks?

2 A load of concrete contains 4.3 kg of sand, 2.8 kg of cement and 1.6 kg of gravel.

   a   What is the total mass of a load of concrete?

   b   How much sand do you need to make 5 loads?

   c   How much cement do you need to make half a load?

   d   How much do 9 loads of concrete weigh?

3 a   Write how much liquid is in each container in litres, as a decimal.

   b   How much liquid must you add to each container, or remove from it, to have the closest whole litre amount?

   c   Calculate 8 times each original amount.

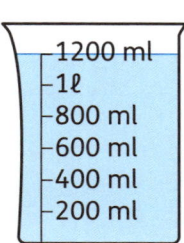 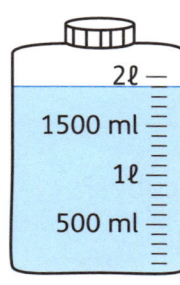 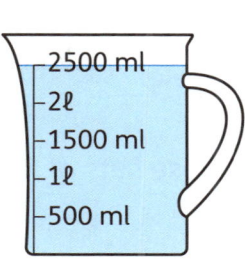 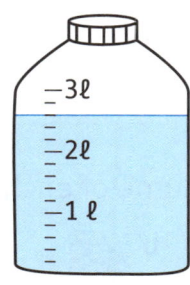 

4 A square has sides of 3.8 m.

   a   What is the perimeter of the square?

   b   Work out the perimeter of a square with sides that are half as long.

   c   Work out the perimeter of a square with sides that are twice as long.

➡ Workbook page 82

Volume and capacity

Volume of 3D objects

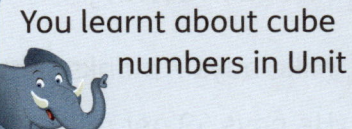

You learnt about cube numbers in Unit 7.

Think and share

 This cube represents 1 cubic centimetre. We write this as 1 cm³.

What do these arrangements of cubes represent?

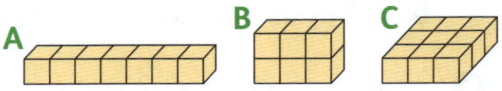

Sometimes you cannot see all the cubes in an arrangement.

How can you work out how many cubes there are in each of these arrangements? Explain your ideas to your group.

Volume is the amount of space that something takes up. We use **cubic units** to describe the volume of solids.

This arrangement of 1 cm³ cubes takes up 4 cubic centimetres of space. It has volume 4 cm³.

This arrangement of 1 cm³ cubes takes up 8 cubic centimetres of space. It has volume 8 cm³.

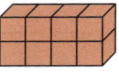

1 What is the volume of each of these sets of 1 cm³ cubes?

a  b c  d  e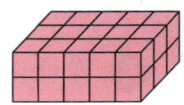

Problem solving

2 This model is made from 1 cm³ cubes. Can you work out its volume?

You can model the shape using cubes.

Think about volume

Look at this arrangement of cubes.

You can think about the shape in different ways.

There are 10 cubes on the top layer.
The shape has 3 layers of 10 cubes.
So, there are 3 layers of 10 cubes.
$3 \times 10 = 30$ cubes

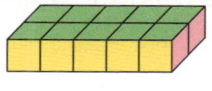
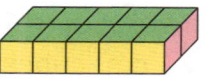

There are 15 cubes on the front layer.
There are 2 layers of 15 cubes.
So, there are $2 \times 15 = 30$ cubes.

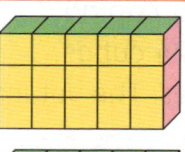
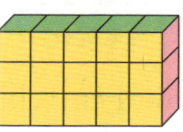

Can you see another way to divide this shape into layers?
Tell your partner.

1 Divide each shape into layers. Write a multiplication to work out the volume of each solid. Each cube is 1 cm³.

a

b

c

d

e

f

g

➡ *Workbook page 83*

Find the volume of containers

How many more of the smaller boxes do you think will fit into this box? Give reasons.

We can work out how much a box can hold by working out how many 1 cm³ cubes will fit inside it. This is called its volume.

The blue lines represent a box. Some 1 cm³ cubes are shown inside the box. How many 1 cm³ cubes can the box hold?

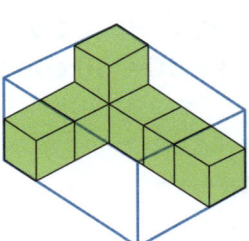

Each layer is a rectangle 3 cubes wide and 4 cubes long.
$3 \times 4 = 12$ cubes.
The box is 2 cubes high. It can hold 2 layers of cubes.
2 layers of 12 = $2 \times 12 = 24$ cubes.
The box can hold 24 cubes. The volume of the box is 24 cm³.

1 The diagrams show some 1 cm³ cubes in different boxes.

  a Work out how many cubes are needed to fill each box.

  b What is the volume of each box, in cubic centimetres?

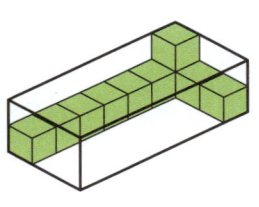 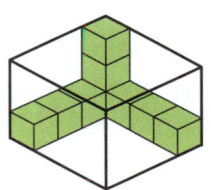  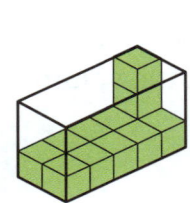 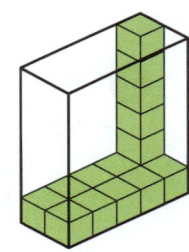

Problem solving

2 Mara wants to fill a cube that has volume 27 cm³.
She has these 1 cm³ cubes. How many more does she need to fill the 27 cm³ cube?

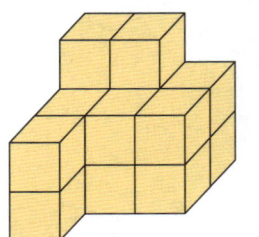

 Model the problem using cubes.

Calculate volume

Solids are 3-dimensional shapes because they have 3 dimensions that we can measure: length, width and height.

Work out the dimensions of cuboid A.

Cuboid A is 3 cubes long, 2 cubes wide and 4 cubes high.

Each cube has edges 1 cm long. Cuboid A has length 3 cm, width 2 cm and height 4 cm.

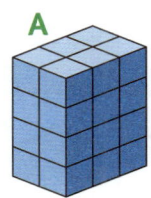

A

Cuboid B has length 8 cm, width 3 cm and height 5 cm. Calculate the volumes of cuboid A and cuboid B.

You cannot count cubes to find the volume of B.

Volume of cuboid = length × width × height

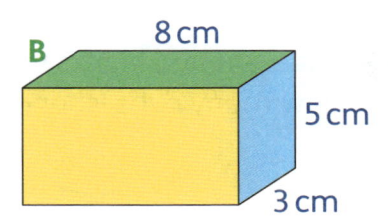

Cuboid	Length in cm	Width in cm	Height in cm	Volume
A	3	2	4	$3 \times 2 \times 4 = 24$ cm³
B	8	3	5	$8 \times 3 \times 5 = 120$ cm³

1 Calculate the volume of each cuboid. The dimensions are in centimetres.

Cuboid	Length	Width	Height
A	10	10	10
B	8	8	8
C	8	7	7
D	20	10	5
E	4	4	4

A cube with edges 1 m long has volume 1 cubic metre, or 1 m³.

2 Here are the dimensions of different shipping containers. Each container is a cuboid. Calculate the volume of each container, in cubic metres.

   a 10 m long, 3 m high and 2 m wide

   b 15 m long, 3 m wide and 3 m high

   c 6 m long, 1.5 m wide and 2 m high

➡ *Workbook page 84*

Solve volume problems

1 A storage company packs boxes into larger containers. Work out how many more boxes can fit into each of these containers.

a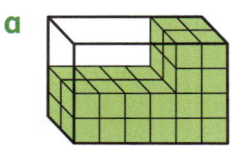

b

c

d

2 Cube-shaped blocks with edges 2 cm long are packed into plastic boxes. How many cube-shaped blocks could fit into the plastic box shown here?

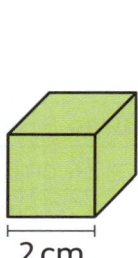
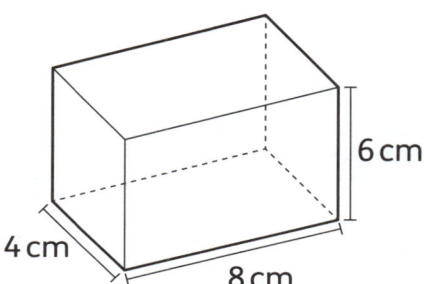

3 How many boxes of cookies can be packed into the big box?

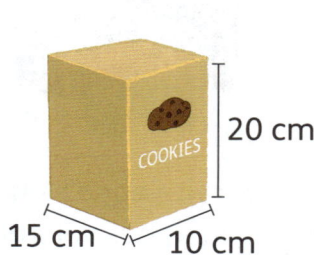

4 This is the floor plan of a hall. Each square on the plan represents 1 m². The walls of the hall are 4 m high. What is the volume of the hall?

Think back to what you learnt about the area of composite shapes.

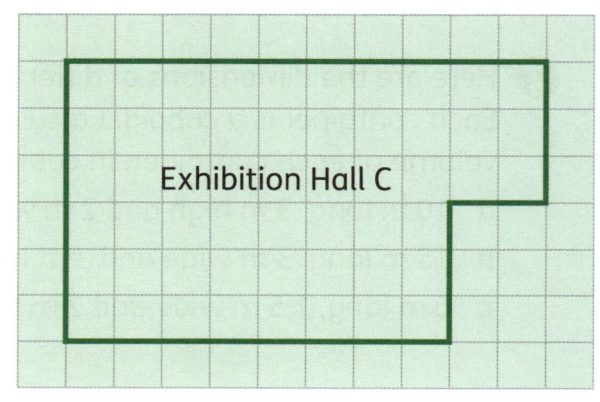

UNIT 17 — Ratio and proportion

What is a ratio?

> ☁️ **Think and share**
>
> A **ratio** is a way of comparing two different amounts part to part.
> Look at this row of hearts and circles:
>
> For every 1 heart there are 3 circles.
> The ratio of hearts to circles is 1 to 3.
> We write this as 1 : 3.
>
> Look at this set of shapes:
>
> Conrad says that the ratio of hearts to circles is 2 to 8.
> Lee says that the ratio of hearts to circles is 1 to 4.
> The teacher says they are both correct. Can you explain why?
>
> Look at this row of shapes:
>
> The ratio of squares to hearts is 2 : 3.
> The ratio of hearts to squares is 3 : 2.
>
> Discuss why the order of numbers in a ratio is important.

1 Write the ratio of stars to planets in each set.

2 What is the ratio of planets to stars in each set?

3 Milo said that the ratio of cars to scooters on the road was 5 : 4.

 a What does this mean?

 b Does a ratio of 5 : 4 mean that Milo only saw 9 vehicles on the road?
 Give reasons for your answer.

What is a proportion?

A **proportion** compares a part to a whole.
Look at this set of hearts and circles.

1 out of every 4 shapes is a heart.
3 out of every 4 shapes are circles.

We can use fractions to write proportions.

1 out of 4 is $\frac{1}{4}$ 3 out of 4 is $\frac{3}{4}$

1 Look at these socks.

   a How many socks are there altogether?

   b What proportion of the socks are plain?

   c What proportion of the socks have stripes?

   d What proportion of the socks are blue?

2 Salma and Marie have these sets of stickers.

Salma's set

Marie's set

Answer these questions for each set. Give your answers as fractions.

   a What proportion of the faces show teeth?

   b What proportion of the faces are smiling?

   c What proportion of the faces are wearing glasses?

   d What proportion of the faces are neither showing teeth, nor smiling nor wearing glasses?

➡ *Workbook page 85*

Work with ratio and proportion

1 We often use ratios in recipes.

Change each set of instructions into a ratio.

a For 1 cup of oats add 2 cups of water.

b Cook 1 cup of rice in 5 cups of water.

c You will need 250 ml of milk per 15 ml of chocolate powder.

2 We also use ratios to help us represent areas on maps.

Write these as ratios. Remember to make sure both numbers in the ratio are expressed in the same units.

a 1 centimetre on the map represents 1 metre in real life.

b 25 cm represents 50 cm.

c 2 cm on the map represents 400 cm.

3 This box contains 6 dark chocolates, 14 milk chocolates and 4 white chocolates.

What proportion of the whole box is:

a dark chocolates

b milk chocolates

c white chocolates?

4 There are 23 sweets in a jar. There are 3 colours: yellow, red and black.

There are 3 times as many yellow sweets as black sweets.

There are 5 times as many yellow sweets as red sweets.

What proportion of the sweets are:

a yellow

b red

c black?

Solve ratio and proportion problems

You can mix paint in different proportions to make new colours.

To make 3 tins of pink paint, you need to mix 2 tins of white paint with 1 tin of red paint.

  +

For every 2 tins of white paint, you need 1 tin of red.

The ratio of white to red is 2 : 1.

2 out of every 3 tins is white.

The proportion of white paint in the mixture is $\frac{2}{3}$. The proportion of red paint in the mixture is $\frac{1}{3}$.

However much pink paint you want, you will always need $\frac{2}{3}$ white and $\frac{1}{3}$ red.

Work with a partner to solve these problems.

1. A painter makes green paint by mixing blue and yellow paint in the ratio 2 : 3. That is 2 tins of blue for every 3 tins of yellow.

   a What proportion of the green paint is blue?

   b If you start with 6 tins of blue paint, how much yellow will you need?

   c If you start with 6 tins of yellow paint, how much blue will you need?

   d If you start with 1 tin of blue paint, how much yellow will you need to make green?

   e How many tins of each colour will you need to make 30 tins of green?

2. To make orange paint, you mix red and yellow paint in the ratio 2 : 5.

   a Make up five ratio and proportion problems using this information.

   b Swap problems with a partner and try to solve each other's problems.

   c Discuss your answers and how you solved each problem.

▶ *Workbook page 86*

Probability

How likely is it?

☁ **Think and share**

What does it mean if we say that something is:

certain   **likely** to happen   **unlikely** to happen   **impossible**?

Decide whether these things are certain, likely, unlikely or impossible.
Explain your choices.

I will get some presents for my birthday.	The Sun will come out tonight.
A dragon will visit my school.	I will go to the Moon one day.
We will have burgers for lunch this week.	There will be 30 days in September this year.

1 Work in groups. Think about your school day. Identify three things that are:

a definitely going to happen (certain)

b definitely not going to happen (impossible).

2 A pupil has two £1 coins, one 50p coin and three 20p coins. Decide whether these statements are true or not.

a It is impossible for the pupil to pick two coins and make 50p.

b If the pupil picks any coin, it will definitely be worth at least 20p.

c If the pupil picks any coin, it is likely to be silver.

d If the pupil picks any coin, it is likely to have a hole in it.

e If the pupil picks two coins, at least one of them will be a 20p coin.

The probability scale

Probability is the chance that something will happen.

We can describe probability using words such as impossible, unlikely, **equally likely**, likely and certain.

Equally likely means that the chance of something happening is the same as the chance of it not happening. For example, if you flip a coin, it is equally likely to land on heads or on tails.

The **probability scale** is a number line with probability words and numbers on it.

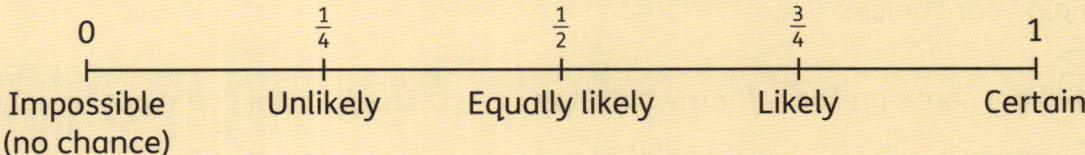

0	$\frac{1}{4}$	$\frac{1}{2}$	$\frac{3}{4}$	1
Impossible (no chance)	Unlikely	Equally likely	Likely	Certain

The scale goes from 0 to 1.
Things that are certain have probability 1.
Things that are impossible have probability 0.
Other things have probability between 0 and 1 based on how likely they are to happen.

1 Describe the chance of each of these things happening. Use words and numbers from the scale.

a You will go to bed early tonight.

b It will rain tomorrow.

c You will be younger tomorrow.

d You will travel by train next week.

e If you flip a coin, it will land on heads.

f It will get dark tonight.

g If you pick a digit card, you will get a number lower than 10.

h In 100 years' time, people will be able to fly.

i Your next lesson will be English.

j There is ice at the North Pole.

2 Ling is carrying a box that contains 3 green balls, 7 white balls and 1 red ball. She trips and one white ball falls out of the box. Write one thing that has probability:

a 0   b 1   c $\frac{3}{5}$

➡ *Workbook page 87*

More probability

If you pick a sock from this drawer without looking, it is likely you will pick a striped sock, unlikely you will pick a spotted sock and impossible you will pick a plain sock.

Probability of striped sock = $\frac{3}{4}$

Probability of spotted sock = $\frac{1}{4}$

Probability of plain sock = 0

1 If you pick one item from each of these sets without looking, write the probability that it is:

   a spotted

   b plain

   c striped.

A   **B**   **C**

2 Here are three spinners. You spin the arrow and it stops on one of the colours.

For each spinner, write the probability of the arrow landing on each colour. Write your answers as fractions.

a   **b**   **c**

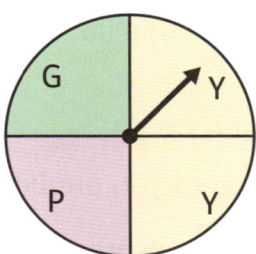
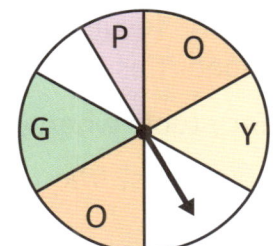

Experiment with probability

When you repeat something a number of times to see how often different things happen, you are doing a **probability experiment**.

For example, you could flip a coin 25 times to see how often it lands heads up.

The **possible outcomes** of a coin are heads or tails. These are the only results you can get.

We use frequency tables to record the outcomes of probability experiments. For the coin-flipping experiment, the table could look like this:

Possible outcome	Tallies	Frequency
Heads	ЖЖ ЖЖ II	12
Tails	ЖЖ ЖЖ III	13
Number of times we repeated the experiment		25

In this experiment, the coin landed heads up 12 times and tails up 13 times.

1. Meg made this spinner and spun it 50 times. She recorded her results using tallies.

   a. Draw a frequency table to organise Meg's results.

   b. Which outcome was most frequent?

   c. Which outcome was least frequent?

2. Work with a partner to do your own probability experiment. You will need a bag and some different coloured counters or cubes.

   a. Make a frequency table to record the possible outcomes. You will need to decide what these are for your set of counters or cubes.

   b. Put a mark next to the colour that you think has the highest probability of being picked.

   c. Put the counters or cubes into the bag. Remove one at a time without looking. Keep a tally of the results. Put the counter or cube back into the bag each time. Repeat this 50 times.

   d. Complete the frequency table when you have finished.

   e. What colour did you pick most frequently? Is that what you predicted? Explain why or why not.

➡ *Workbook page 88*

Mixed practice 3

1 A farmer has 21 rows of peach trees. There are 35 trees in each row. How many trees does she have altogether?

2 There are 19 houses in my road. There are 15 paving stones in front of each house. How many paving stones are in the road?

3 83 × 20 = 1660. Show how you can use this fact to work out these calculations without doing any multiplication or division.

    a 21 × 83   **b** 19 × 83   **c** 1660 ÷ 20   **d** 1660 ÷ 83

4 An elevator at Burj Khalifa, the tallest building in the world, can carry 25 passengers at a time. The elevator went up to the top of the building 42 times in one day. What is the maximum number of passengers it could have carried on that day?

5 Jess wants to divide 800 by 25.

This is what she writes in her book:

> 800 ÷ 25
> 800 ÷ 100 = 8
> 8 × 4 = 32

    a Is her answer correct?

    b Show how she could set out her working more clearly.

6 Look at this line graph carefully.

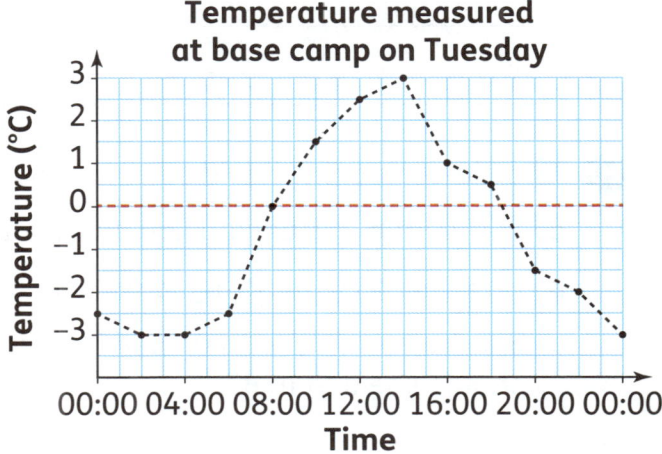

Temperature measured at base camp on Tuesday

    a What does the red dashed line across the graph show?

    b What is the highest temperature measured in this period?

    c What was the coldest temperature measured on Tuesday?

    d Between which two times did the temperature rise the fastest? How do you know this?

    e Estimate the temperature at 3 p.m.

    f At what times was the temperature −2.5 °C?

    g What is the temperature 4 degrees lower than −3 °C?

7 This diagram has been drawn on a grid. The squares on the grid have sides 1 cm long.

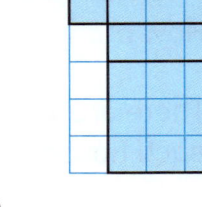

    a What do we call this type of diagram?

    b Draw the 3D shape you could make from this diagram. Write the length, width and height on the shape.

    c Calculate the volume of the 3D shape.

    d How many cuboid-shaped erasers 2 cm long, 1 cm wide and 1 cm high could you fit into a box this size?

8 Round each decimal to the nearest whole number.

Estimate the answer then complete each calculation.

    a $3.63 + 6.21$ **b** $98.76 - 54.12$ **c** 21.4×10

    d 8.6×4 **e** $18.45 - 12.32$ **f** $54.6 \div 10$

9 Two pupils were asked to add 0.55, 11.134 and 4.9.

This is how they set out their work.

Pam	Sam
0 · 5 5	5 5
1 1 · 1 3 4	1 1 1 3 4
+ 4 · 9	+ 4 9

    a Who will get the correct answer?

    b What mistakes has the other pupil made?

10 A bottle of cleaning liquid says: Add 50 ml per litre of water.

    a Is the ratio of cleaning liquid to water 50 : 1? Give a reason.

    b How much cleaning liquid will you add to 5 litres of water?

    c Sondra has only got 20 ml of cleaning liquid left. How much water should she add to it?

11 Water flows into a tank at a rate of 5 litres every 20 seconds.

    a What is the rate of water flow in litres/minute?

    b How much water will flow into the tank in half an hour?

Glossary

A

acute angle – an angle between 0° and 90°

angle – formed when two lines meet at a point (vertex); the size of an angle is the amount of turn from one of the lines to the other, measured in degrees

area – the amount of space covered by or inside a 2D shape, measured in square units such as cm² and km²

ascending – arranged in order from smallest to greatest

axis (plural **axes**) – the labelled horizontal and vertical lines that form the framework for a graph

B

bar chart – a graph that uses horizontal or vertical bars to show different categories of data; the lengths of the bars show frequencies

bar-line chart – a bar chart where the bars are replaced with horizontal or vertical lines that show the frequencies

C

capacity – the amount of liquid a container can hold, measured in units such as millilitres (ml) and litres (ℓ)

Carroll diagram – a table for sorting items according to two sets of categories

certain – a probability word meaning something will definitely happen

common factor – a whole number that can divide (without remainder) into two or more numbers; for example 2 is a common factor of 4 and 6

common multiple – a multiple that is shared by two or more numbers; for example 12 and 24 are common multiples of 3 and 4

composite number – a whole number greater than 1 with more than two factors

composite shape – a shape made by combining two or more other shapes

coordinates – a pair of numbers that gives the position of a point on a grid, written in brackets with x-coordinate, then y-coordinate; for example (2, 4)

coordinate grid – a grid formed by a pair of numbered axes; the x-axis is horizontal and the y-axis is vertical

cube number – the product when a number is multiplied by itself three times; for example 8 is a cube number ($2 \times 2 \times 2$)

cubic unit – unit of volume, for example centimetre cubed or cm³

D

decimal – a decimal fraction with digits to the right of the decimal point, representing tenths, hundredths, thousandths and so on, for example 0.2, 0.05, 0.375

denominator – the bottom number in a fraction; for example 3 is the denominator of $\frac{1}{3}$; the denominator tells you how many parts the whole is divided into

descending – arranged in order from greatest to smallest

dimension – a measurement such as length, width or height

dot plot – a graph that shows data as dots

E

equally likely – outcomes with the same or even chance of happening

equilateral triangle – a regular three-sided polygon; a triangle with equal angles and equal side lengths

equivalent fractions – fractions with the same value, for example $\frac{1}{2}$ is equivalent to $\frac{4}{8}$; fractions can also be equivalent to decimals and percentages, for example $\frac{1}{2}$, 0.5 and 50% are equivalent

F

factor – a whole number that divides exactly into another number without leaving any remainder; for example 1, 2, 3 and 6 are all factors of 6

formula – a rule written in words or using letters; for example Area of rectangle = length × width, or $A = l \times w$

frequency table – a table showing how many times a value or event occurs

H

heptagon – a polygon with seven sides

highest common factor (HCF) – the greatest number in the set of common factors of two or more numbers

I

imperial – an older system of measurement with non-decimal units; inches, feet, miles, ounces, pounds and pints are imperial units

improper fraction – a fraction in which the numerator is greater than (or equal to) the denominator; for example $\frac{5}{3}$

impossible – a probability word that means something cannot happen; no chance

inverse operation – an operation that undoes another operation; addition and subtraction are inverse operations, for example, 2 + 3 = 5 and 5 – 3 = 2; multiplication and division are inverse operations

isosceles triangle – a triangle with two sides of equal length and two equal angles

L

likely – a probability word that means that there is a greater than even chance of something happening

line graph – a graph made of line segments that join points to show patterns or changes in the data

lowest common multiple (LCM) – the smallest number in the set of common multiples of two or more numbers

M

median – the middle value in an ordered set of data; for example the median of the data set 1, 2, 2, 9, 10, 12, 13 is 9

metric system – an international system of standard units of measure with decimal units; centimetres, metres, kilometres, grams, kilograms, millilitres and litres are metric units

million – one thousand thousand, 1 000 000

mirror line – line of symmetry, the line in which a shape is reflected

mixed number – a number with a whole number part and a fraction part; for example $1\frac{2}{3}$

mode – the value that appears most often in a set of data; for example the mode of the data set 1, 2, 2, 9, 10, 12, 13 is 2

multiple – a multiple of a number is the product of that number and a whole number; for example 3, 6, 9 and 12 are all multiples of 3

N

negative number – a number less than zero, written with a minus sign; for example –2

net – a 2D shape that folds to make a 3D shape

nonagon – a polygon with nine sides

numerator – the top number in a fraction; for example 3 is the numerator in $\frac{3}{8}$; the numerator tells you how many parts of the whole you are dealing with

O

obtuse angle – an angle greater than 90° but less than 180°

P

percentage – a fraction expressed as a number or amount out of 100, using the % sign; for example $\frac{1}{2}$ or $\frac{50}{100}$ is 50%

perimeter – the distance around the outside of a 2D shape

place value – the value of each digit in a number, for example, in 2641 the '6' has a place value of '6 hundreds'

polygon – a 2D shape with straight sides

positive number – a number greater than zero

possible outcomes – all of the possible results of an experiment; for example if you flip a coin, there are two possible outcomes: heads or tails

power of 10 – the product when 10 is multiplied by itself a certain number of times; for example 10 to the power of 2 is 10 × 10 or 100; the first six powers of 10 are 10, 100, 1000, 10 000, 100 000 and 1 000 000

prime number – a whole number with exactly two factors: 1 and the number itself; 2, 3, 5, 7 are prime numbers

probability – a measure of the chance that something will happen, written in words, or as a value between 0 and 1 or as a percentage

probability experiment – a number of tests or trials used to investigate how often an outcome occurs

probability scale – a scale from 0 to 1 used to give a value to the probability of something happening, where 0 means an outcome is impossible and 1 means it is certain

proportion – a part compared to a whole; can be given as a fraction or percentage; for example, if there are 10 marbles and 3 or them are blue, $\frac{3}{10}$ of the marbles are blue

R

range – the difference between the greatest and smallest values in a set of data; for example, for the data set 1, 2, 2, 9, 10, 12, 13 the range is 13 − 1 = 12

rate – compares two different quantities; for example, speed in km per hour is the rate of change of distance with time

ratio – a comparison of two amounts in a particular order; for example, for a drink made of 1 part orange and 5 parts water, the ratio orange : water = 1 : 5

reflection – a mirror image or when something is flipped in a mirror line

reflex angle – an angle greater than 180° but less than 360°

remainder – the amount left over after dividing into equal groups, 5 divided by 2 makes two equal groups of 2 and leaves a remainder of 1, or 2 r 1

round (a number) – write a number with zeros in place of some digits to make it easier to work with; decimal numbers can be rounded to the nearest whole number, tenth, hundredth and so on

S

scalene triangle – a triangle with no equal sides or angles

sequence – a number pattern or shape pattern that follows a rule to get from one term to the next

short division – a written method of division by a 1-digit number

square number – the product of a number and itself; for example 4 × 4 = 16, so 16 is a square number

square unit – unit of area, for example square centimetres (cm^2)

squared – a number is squared when it is multiplied by itself

symmetrical – a shape (or pattern) is symmetrical if it can be folded (reflected) along a line to form two equal parts that are mirror images of each other

T

terms – the individual numbers or patterns in a sequence

term-to-term rule – the rule that describes how to get from one term in a sequence to the next

thousandth – when you share 1 whole into 1000 equal parts, each part is one-thousandth; 1 thousandth = $\frac{1}{1000}$ = 0.001

tonne – a metric measure of mass equivalent to 1000 kilograms

translation – a sliding movement right, left, up or down so that all points on the shape move the same distance in the same direction

trapezium – a quadrilateral with one pair of parallel sides

U

unlikely – a probability word that means there is a lower than even chance of something happening

V

Venn diagram – a diagram that uses circles to sort objects, shapes and numbers into sets; items written where the circles overlap are in both sets

vertex (plural **vertices**) – the point where two (or more) lines meet or intersect; a corner of a polygon or 3D shape

volume – the amount of space occupied by a 3D object; volume is measured in cubic units, for example centimetres cubed (cm^3)

W

whole – all the parts; the total amount